<u>Disclaimer</u>

Book Title: Performance of New and Aged Residential Fire Sprinklers

Book Author: Anthony D. Putorti Jr; William F. Guthrie; Jason D. Averill; Richard G. Gann;

Book Abstract: The U.S. Consumer Product Safety Commission (CPSC) initiated a program to determine the effects of emissions from problem drywall on residential electrical, gas distribution, and fire safety components. As part of this program, the National Institute of Standards and Technology (NIST) generated data to help determine whether there has been degradation in the activation performance of automatic residential fire sprinklers exposed to those emissions, as manifested by changes to sprinkler activation time. NIST tested three sets of sprinklers in the sensitivity test oven (plunge test apparatus), according to the oven heat test section of UL 199 / UL 1626. Set 1 (bulb type) residential sprinklers were provided by CPSC staff and described as having been installed in homes with problem drywall; Set 2 sprinklers (bulb and fusible types) were purchased new by NIST and tested as received; and Set 3 comprised new sprinklers, of the same models as Set 2, after they had been subjected to an accelerated aging protocol, the Battelle Class IV corrosivity environment. Sprinklers from all three sets were installed and tested in the UL 199 / UL 1626 plunge test apparatus.

Citation: NIST TN - 1702

Keywords: drywall; fire; fire sprinkler; sprinkler; UL 199; UL 1626; plunge test

PERFORMANCE OF NEW AND AGED RESIDENTIAL FIRE SPRINKLERS

Anthony D. Putorti, Jr.

Jason D. Averill

Richard G. Gann

Fire Research Division
Engineering Laboratory

William F. Guthrie

Statistical Engineering Division
Information Technology Laboratory

U.S. Department of Commerce

Rebecca M. Blank, Acting Secretary

Patrick D. Gallagher

Under Secretary of Commerce for Standards and Technology
Director, National Institute of Standards and Technology

National Institute of Standards and Technology Technical Note 1702

Natl. Inst. Stand. Technol. Tech. Note 1702, 92 pages (August 2011)

CODEN: NTNOEF

Technical Report to the U.S. Consumer Product Safety Commission

CPSC-I-09-0024

Rohit Khanna, Fire Protection Engineer

This report was prepared for the Commission pursuant to Interagency Agreement CPSC-I-09-0024. It has not been reviewed or approved by, and may not necessarily reflect the views of, the Commission.

This page intentionally left blank.

Disclaimer

Certain commercial equipment, instruments, materials, standards, or laboratories are mentioned in the text to specify the experimental procedure and equipment adequately. In no case does such identification imply recommendation or endorsement by the National Institute of Standards and Technology, nor does it imply that the equipment, materials, or standards are the best available for the purpose.

Regarding non-metric units: The policy of the National Institute of Standards and Technology is to use metric units in all of its published materials. To aid the understanding of this report, in most cases, measurements are reported in both metric and U.S. customary units.

This page intentionally left blank.

ABSTRACT

The U.S. Consumer Product Safety Commission (CPSC) initiated a program to determine the effects of emissions from problem drywall on residential electrical, gas distribution, and fire safety components. As part of this program, the National Institute of Standards and Technology (NIST) generated data to help determine whether there has been degradation in the activation performance of automatic residential fire sprinklers exposed to those emissions, as manifested by changes to sprinkler activation time. NIST tested three sets of sprinklers in the sensitivity test oven (plunge test apparatus), according to the oven heat test section of Underwriters Laboratories (UL) UL 199 / UL 1626. Set 1 (bulb-type) residential sprinklers were provided by CPSC staff and described as having been installed in homes with problem drywall; Set 2 sprinklers (bulb and fusible-types) were purchased new by NIST and tested as received; and Set 3 was comprised of new sprinklers, of the same models as Set 2, which had been subjected to a simulated 20-year exposure to problem drywall emissions by means of an accelerated aging protocol, the Battelle Class IV corrosivity environment. Sprinklers from all three sets were installed and tested in the UL 199 / UL 1626 plunge test apparatus.

NIST found that one of the sprinklers, a fusible-type subjected to the accelerated aging protocol, did not activate during testing. All other tested samples of fusible-type sprinklers, new (Set 2) and aged (Set 3), activated within the maximum activation time specified by UL 199 in the plunge test apparatus. All of the bulb-type sprinklers activated within the maximum activation time specified by UL 1626 in the plunge test apparatus. NIST found a statistically significant difference in activation time for one model of bulb-type sprinkler between Set 2 and Set 3. This difference in activation time, while statistically significant, was numerically small, compared to the maximum specified activation time. There were no statistically significant differences in activation times between sprinklers in Set 1, and similar sprinklers from Set 2.

Keywords: drywall; fire; fire sprinkler; sprinkler; UL 199; UL 1626; plunge test

This page intentionally left blank.

SUMMARY

The U.S. Consumer Product Safety Commission (CPSC) has established an engineering test program to determine the effects of emissions from problem[i] drywall on residential electrical, gas distribution, and fire safety components as part of its overall investigation. As part of this program, under an agreement with the CPSC, the National Institute of Standards and Technology (NIST) has conducted testing to help determine whether there has been degradation in activation performance of automatic fire sprinklers (hereafter referred to as "sprinklers") exposed to those emissions, as manifested by changes to sprinkler activation time. This document presents the results of the NIST testing of sprinklers under the agreement.

The testing was designed to address three questions:

1. Do sprinklers exposed to problem drywall emissions activate within allowable time specifications?

2. Do sprinklers exposed to problem drywall emissions activate differently from new sprinklers of the same model?

3. Does a 20-year exposure to problem drywall emissions, simulated by a four-week exposure to the Battelle Class IV corrosivity environment using an accelerated aging test protocol, affect sprinkler activation time performance?

NIST conducted tests of 81 sprinklers in the sensitivity test oven (plunge test apparatus), according to the oven heat test section of Underwriters Laboratories (UL) 199 / UL 1626. UL 199 is a national standard for automatic fire sprinklers, while UL 1626 is a national standard for automatic residential fire sprinklers.[ii] The standards are discussed in *Section IV.A* of this report.

The sprinklers were categorized into three sets:

(Set 1) Nine sprinklers, provided by CPSC staff and described as having been installed in homes constructed around 2006 with problem drywall, and removed in 2009;

(Set 2) Thirty-six new sprinklers purchased by NIST in 2010, tested as received; and

(Set 3) Thirty-six of the same sprinkler models[iii] as Set 2, which were subsequently exposed, through an accelerated aging protocol, to a concentration of corrosive gases designed to simulate a 20-year exposure.

---

[i] CPSC uses the term "problem drywall" to refer to drywall associated with corrosion of metal in homes. This has also been reported as "Chinese drywall," but CPSC staff has found that not all Chinese or imported drywall is problem drywall, and has not ruled out the possibility that some domestic drywall could be problem drywall. For additional information see: http://www.cpsc.gov/library/foia/foia11/os/staffdomdrywall2011.pdf.

[ii] Model building codes and the National Fire Protection Association (NFPA) codes require that only sprinklers that have been evaluated in accordance with the applicable standard and listed by an approved certification laboratory are acceptable for installation. One of the requirements for listing is passing the test(s) related to maximum allowable activation time. Nearly all sprinklers available for purchase in the United States are listed by UL. There are other listing/approval organizations that publish their own test standards, however, such as FM Global (FM), Norwood, MA. Many sprinkler models are listed / approved by UL and FM. All of the sprinklers in this study are listed by UL. Two of the six sprinkler models also carried FM approvals, both of which were fusible-type quick response (QR) sprinklers. NIST does not endorse any particular listing or approval organization.

[iii] NIST purchased six different sprinkler models, with equal quantities of each model populating Set 2 and Set 3. Each of the sprinkler models was purchased from the same vendor at the same time and arrived in the same box as packaged by the manufacturer.

All of the Set 1 sprinklers provided by the CPSC were the pendent style, equipped with bulb-type operating elements. Equal numbers of the sprinklers purchased from fire equipment vendors in Sets 2 and 3 were equipped with bulb-type and fusible-type, heat activated operating elements. The purchased sprinklers included pendent and sidewall styles and had a mix of white, chrome, and brass/bronze finishes.

NIST found that one sprinkler, a fusible-type subjected to the accelerated aging protocol, did not activate during testing. All other tested samples of fusible-type sprinklers, new (Set 2) and aged (Set 3), activated within the maximum activation time specified by UL 199 in the plunge test apparatus. All of the bulb-type sprinklers activated within the maximum activation time specified by UL 1626 in the plunge test apparatus.

For all three models of fusible-type sprinklers, and for two of the three models of bulb-type sprinklers, the data showed no statistically significant effect of the accelerated aging protocol on activation time. NIST found a statistically significant difference in activation time for one model of bulb-type sprinkler between the units in Set 2 (new) and the units in Set 3 (aged). This difference in activation time, while statistically significant, was numerically small compared to the maximum specified activation time.

No statistically significant differences in activation times were found between sprinklers in Set 1 (recovered from homes) and similar sprinklers from Set 2 (new). This indicates that for the bulb sprinklers in Set 1, the data showed no statistically significant effect on activation time from the approximate 3-year exposure to problem drywall emissions.

Table of Contents

This page intentionally left blank.

I. OBJECTIVE AND BACKGROUND

At the request of the U.S. Consumer Product Safety Commission (CPSC), the National Institute of Standards and Technology (NIST) tested and analyzed the performance of automatic fire sprinklers (hereafter referred to as "sprinklers") to aid the CPSC in ascertaining possible changes in functionality and resulting fire safety risk arising from sprinkler exposure to emissions from problem drywall. This Technical Note documents the NIST research. A companion study examined the performance of smoke alarms as part of the CPSC study [1].

Some problem drywall installed in U.S. homes has been reported to be associated with corrosion to central air conditioning components, copper tubing, and exposed copper wiring [2]. There have been reports of premature failures of various household items: air conditioner evaporator coils, electric appliances, televisions, and electrical switches in homes. A range of health issues have been reported by residents in these homes as well [3].

The CPSC's investigation into the possible effects of problem drywall includes three tracks: (1) evaluating the relationship between the drywall and reported health issues; (2) evaluating the relationship between the drywall and possible effects on electrical, gas distribution, and fire safety components that could result in potential fire and shock hazards; and (3) tracing the origin and distribution of the drywall in commerce to identify the scope of potential problems presented by the drywall [2].

As part of this investigation, the CPSC and its partners conducted the following studies to determine the differences in composition between problem drywall and typical domestic drywall:

- In-home air sampling studies, including continuous, real-time measurements of sulfur compounds, acids, and other gases. The sampling extended over days because many health symptoms reportedly occurred after hours of sleeping [2].

- Laboratory elemental analysis of typical domestic and problem drywall, with characterization of any differences.

- Laboratory chamber studies of typical domestic and problem drywall to separate and isolate chemical emissions from drywall, as opposed to chemicals emitted from other home products (*e.g.*, carpets, cleaners, paint, adhesives, and beauty products).

These studies have enabled the CPSC and its partners to identify problem drywall and to characterize differences between the chemical composition of gases emitted by problem drywall and gases emitted by typical domestic drywall [4].

A reference environment has been established for replicating the effects of the emissions on metallic and electrical components and, by increasing the temperature and concentration of the contaminating gases, accelerating the aging of the components [5]. This environment, designated Battelle Level IV, is an accelerated aging protocol designed to replicate the outcome of 20 years of exposure in four weeks [6].

This report describes the operating principles of the sprinklers, characterizes the sprinklers included in the NIST evaluation, details the apparatus and procedures used to measure sprinkler performance, and presents the test data. The report concludes with an analysis of the test results with regard to requirements for sprinkler activation time, as well as the degree of change in activation time as a result of exposure to contaminated atmospheres.

This page intentionally left blank.

II. FIRE SPRINKLER ACTIVATION PRINCIPLES

A. FIRE SPRINKLER ACTIVATION

An automatic fire sprinkler system can be effective in controlling the growth and spread of flaming fires in structures. Typically, as a flaming fire grows, a buoyant plume of heated combustion products and entrained gases rises, eventually spreading along the ceiling and walls. As the heated gases flow past sprinkler(s) mounted near ceilings or walls, the sprinklers absorb heat and may warm to a temperature sufficient for activation. Upon activation, water flows through the sprinkler(s) and directly suppresses the fire and/or cools the temperature in the room as the water droplets change phase from liquid to vapor. Thus, the time to activation of a sprinkler exposed to a specific thermal environment, defined primarily by temperature and gas velocity, is a critical parameter in determining the fire safety effectiveness of a sprinkler system. This time to activation is the focus of the current study.

Model building codes in the United States [7,8] require fire sprinklers to be installed in most rooms of new homes to sense the presence of an unwanted fire rapidly and accurately, and to provide water spray to control the growth of the fire. (State and / or local jurisdictions, however, may or may not adopt these model code provisions.) Residential sprinkler systems are designed to prevent flashover and provide additional time for occupants to exit the structure. The model codes also require that these sprinklers be listed.[iv] Typically, residential sprinklers are listed by Underwriters Laboratories (UL), which requires adherence to UL 1626 [9] test requirements to be deemed acceptable for installation in residences. The oven heat test (plunge test) is one of the test methods contained in UL 1626. The plunge test is used to characterize the response time of sprinklers, which are commonly based on one of two types of operating elements: bulb and fusible.

B. BULB-TYPE FIRE SPRINKLERS

Bulb-type sprinklers contain a heat-actuated element consisting of a frangible glass bulb filled with liquid and a small bubble of gas. Under normal conditions, the bulb provides a force necessary to hold a valve cap against the sprinkler orifice, thereby preventing water flow. Under fire conditions, hot fire gases travel past the operating element as they rise above the fire and move across ceilings and along walls. Heat transferred to the sprinkler from the fire and hot gases causes the temperature of the bulb to rise. When the liquid in the bulb is heated, it expands, raising the pressure within the bulb. As the volume of the gas bubble within the bulb approaches zero, the pressure rises rapidly, causing the bulb to shatter. The shattering of the glass bulb releases the valve cap, allowing it to be forced open by the water pressure in the sprinkler. Water exits the sprinkler orifice, impacts a strike plate, and is delivered as water spray to the room. Each sprinkler of the sprinkler system activates independently.

[iv] Listing of fire sprinklers is required by model building codes and NFPA codes. (Listing is a type of certification that involves determining that the sprinklers meet the performance requirements of the certification laboratory, review of manufacturer quality control procedures, and follow-up audits of sprinkler performance and manufacturing facilities.) One of the requirements for listing is passing the test(s) related to minimum activation time. Nearly all sprinklers available for purchase in the United States are listed by UL. There are other listing/approval organizations that publish their own test standards, however, such as FM Global, Norwood, MA. Many sprinkler models are listed/approved by both UL and FM. NIST does not endorse any particular listing or approval organization.

The bulb-type sprinklers tested in this study were all pendent or sidewall-style residential sprinklers, with temperature ratings of 68 °C (155 °F). The purchased sprinklers included white, chrome, and brass finishes. Recessed, flush, and concealed styles were not tested.

C. FUSIBLE-TYPE FIRE SPRINKLERS

Fusible-type sprinklers contain a heat-actuated element incorporating a metal alloy that changes phase (fuses) when heated above a characteristic temperature. Under normal conditions, the detection element applies restraining forces to a mechanism that holds the valve cap against the sprinkler orifice, preventing the passage of water. Under fire conditions, heat transferred to the sprinkler from the fire and hot gases causes the temperature of the fusible element to rise. When the element reaches its characteristic temperature (fusion temperature), and sufficient energy is transferred to fuse the element, the sprinkler mechanism is released. The release of the mechanism causes the release of the valve cap, which is forced open by the water pressure in the sprinkler. Water exits the sprinkler orifice, impacts the strike plate, and is delivered as water spray to the room and its contents. Each sprinkler of the sprinkler system activates independently.

The fusible-type sprinklers tested in this study were all pendent[v] or sidewall[vi] style quick response (QR) sprinklers, with temperature ratings of 74 °C (165 °F). The purchased sprinklers included a mix of white, chrome, and bronze finishes. Recessed, flush, and concealed styles were not tested.

Conventional fusible-type pendent and sidewall residential sprinklers were not readily available at the time of this study. In order to study the effects of problem drywall emissions on fusible sprinklers, nonresidential QR sprinklers were purchased that have similar activation time requirements as residential sprinklers. Nonresidential sprinklers, as part of a complete automatic sprinkler system, are used to increase life safety in light hazard building occupancies such as educational facilities, hotel sleeping rooms, board and care facilities, hospitals, and nursing homes. [10]. In addition, NFPA 13D [11] and NFPA 13R [12] allow the use of nonresidential, QR sprinklers in some areas of residential structures.

---

[v] A pendent sprinkler is designed to be installed such that the threaded base of the sprinkler is at the top and the deflector is at the bottom. The water stream from the sprinkler orifice is directed downward toward the deflector. See NFPA 13 for more details.

[vi] A sidewall sprinkler is designed to be mounted horizontally along the vertical surface of a wall. They are equipped with special deflectors that are designed to discharge most of the water away from the adjacent wall, in a one quarter sphere pattern. See NFPA 13 for more details.

III. EXPERIMENTAL PLAN

A. GENERAL FORMULATION

The experimental plan was designed to address three questions:

1. Do sprinklers exposed to problem drywall emissions activate within allowable time specifications?

2. Do sprinklers exposed to problem drywall emissions activate differently from new sprinklers of the same model?

3. Does a 20-year exposure to problem drywall emissions, simulated by a four-week exposure to the Battelle Class IV corrosivity environment using an accelerated aging test protocol, affect sprinkler activation performance?[vii]

In order to answer these questions, the exposure history of the sprinkler was a critical consideration. Thus, the plan included testing of sprinklers from three exposure conditions:

(Set 1) Nine sprinklers, provided by CPSC staff and described as having been installed in homes constructed around 2006 with problem drywall, and removed in 2009;

(Set 2) Thirty-six new sprinklers purchased by NIST in 2010, tested as received; and

(Set 3) Thirty-six of the same sprinkler models[viii] as Set 2 that had been subjected, through an accelerated aging procedure, to a concentration of Battelle Level IV corrosive gases designed to simulate 20-year exposure to the emissions from problem drywall.

Additional important details in the experimental plan include the following:

- Sets 2 and 3 included both bulb-type and fusible-type sprinklers. Bulb-type sprinklers were reported by the CPSC as the only sprinklers available for collection by the CPSC staff from problem drywall homes (Set 1).

- While CPSC did not provide any fire sprinklers that were represented to be from homes constructed around 2006 without problem drywall, one of the sprinkler models purchased by NIST (Manufacturer 3, bulb-type) was the same as the sprinkler model in Set 1, albeit with a different finish and manufacturing date.

- Devices from more than one manufacturer were tested, where possible, to ensure the scope of the study included potentially relevant design differences. Differences between sprinklers obtained from different manufacturers also included sprinkler styles, pendent and sidewall, and three different sprinkler coatings: white, chrome, and brass/bronze.

- Replicate tests were performed to enable estimation of experimental repeatability.

[vii] NIST did not investigate the appropriateness of the composition of this particular environment, nor did NIST investigate whether four weeks of exposure in this environment was equivalent to 20 years of exposure of sprinklers to the emissions from problem drywall.

[viii] NIST purchased six different sprinkler models, with equal quantities of each model populating Set 2 and Set 3. Each of the sprinkler models was purchased from the same vendor at the same time and arrived in the same box as packaged by the manufacturer.

Tests to determine sprinkler activation times were conducted at NIST using the sensitivity test oven (plunge test) apparatus, described in UL 199 / UL 1626, which exposes the sprinkler to a stream of heated air. A description of the principles of operation of the plunge test apparatus appears in *Section IV*.

B. FIRE SPRINKLER MODELS

Seven different models of sprinkler were tested. Under ideal circumstances, the same models of sprinkler, from the same manufacturing batches, would be evaluated as part of all three sets. However, in the time between sprinkler installation in homes with problem drywall (generally 2006), and the time that sprinkler evaluation was initiated at NIST (2010), sprinklers from the same manufacturing batch could no longer be purchased as new units. Furthermore, for sprinkler models that continued to be available from 2006 to 2010, it is unknown what changes were made to the model in the intervening years.

The sprinkler models that populate the three Sets listed in Section III.A of this Technical Note are described further as follows:

1. CPSC staff provided nine sprinklers described to be from homes constructed around 2006 with problem drywall, comprising one model (denoted as Set 1 or "field" in this report).[ix] All nine units were bulb-type, pendent residential sprinklers, with nominal temperature ratings of 68 °C (155 °F), and a white coating on the sprinkler body. They were marked with manufacture dates of 2005 and 2006. CPSC did not provide any fusible-type sprinklers because none were found in the sampled homes; nor did CPSC represent that any of the sprinklers provided were from nonproblem drywall homes. All of the sprinklers provided by CPSC were tested in the plunge test apparatus.

2. NIST purchased from fire equipment vendors 150 new bulb-type residential sprinklers, with nominal temperature ratings of 68 °C (155 °F), comprising 50 of each of three different models.[x] NIST also purchased, from fire equipment vendors, 150 new fusible-type QR sprinklers, with nominal temperature ratings of 74 °C (165 °F), comprising 50 of each of three different models. The sprinklers were all produced by three manufacturers, with one model of bulb-type sprinkler and one model of fusible-type sprinkler from each of the three manufacturers. The purchased sprinklers included pendent and sidewall styles, as well as white, chrome, and brass/bronze finishes. Six new units of each model (for a total of 36) were tested in the plunge test apparatus as received. Note that one of the bulb-type

---

[ix] For an ideal comparison of the effects of exposure to problem drywall and exposure to nonproblem drywall, sprinklers would be available from homes where all the other exposure conditions were identical. These include, but are not limited to, criteria such as the sets of homes being otherwise identical, the sprinklers being from the same batches, the indoor contaminant levels in the homes being otherwise identical, sprinklers having been installed at the same time and in similar locations. This was not possible in this study due to the lack of sprinklers from nonproblem drywall homes. This limited, but did not eliminate, the ability to derive information regarding the effects of exposure to the emissions from problem drywall.

[x] To enable comparison between the sprinklers in Sets 1, 2, and 3, it would have been preferable for the sprinklers in Sets 1, 2, and 3 to be of the same model and from the same manufacturing lot. In 2010, the sprinklers in Set 1 were approximately 4 years old. Thus, it was not possible to purchase new units from the same batches. Furthermore, in the ensuing 4 years, model updates may have occurred. This limited, but did not eliminate, the ability to derive information regarding the effect of the nominal 3-year exposure in Set 1, newly manufactured sprinklers in Set 2, and the simulated 20-year exposure in Set 3.

sprinkler models purchased by NIST was the same manufacturer (Manufacturer 3) and the same model number as the sprinklers provided by the CPSC but with a different finish and year of manufacture.

3. NIST shipped 36 of the purchased sprinklers described above (six of each model) to Sandia National Laboratories for exposure to an accelerated aging protocol. After the accelerated aging exposure, the sprinklers were shipped back to NIST, where they were all tested in the plunge test apparatus.

A summary of the number of sprinklers of each type that were tested, and some of their other associated properties, is shown in Table 1. Other than the exposure of each sprinkler, denoted by their designations into sets, and the day of each test (not shown in this table), the other sprinkler properties shown here were not actually treated as experimental factors whose effects are directly assessed in the statistical analysis. However, as discussed in Appendix A, these properties (sprinkler element type, manufacturer, design, and finish) are used to define the populations[xi] within which the effects of the primary factors of interest (exposure and day of testing) are compared. Thus, the scope of the overall study addresses sprinkler properties of secondary interest through the parallel analyses carried out for each set of data.

Exposure Condition	Sprinkler Element Type	Design	Finish	Mfr.	Number of Units	Number of Activation Tests
Set 1 — Approximate 3-year Exposure to Problem Drywall (Provided by CPSC)	Bulb	Pendent	White	3	9	9
Set 2 — New Sprinklers from Commercial Sources	Bulb	Pendent	Brass	1	6	6
		Sidewall	White	2	6	6
		Pendent	Chrome	3	6	6
	Fusible	Pendent	White	1	6	6
		Sidewall	Chrome	2	6	6
		Pendent	Bronze	3	6	6
Set 3 — New Sprinklers with Accelerated 20-year Exposure to Battelle Class IV Corrosion (Accelerated Aging)	Bulb	Pendent	Brass	1	6	6
		Sidewall	White	2	6	6
		Pendent	Chrome	3	6	6
	Fusible	Pendent	White	1	6	6
		Sidewall	Chrome	2	6	6
		Pendent	Bronze	3	6	6

TABLE 1: TEST MATRIX

[xi] Although these factors define the different populations of sprinklers of interest, the sprinklers obtained for testing were not randomly sampled from these populations. Therefore, the results presented here apply primarily to the sprinklers actually sampled, and apply to the larger populations only to the extent that the sampled sprinklers are representative of their respective populations.

C. ACCELERATED AGING PROTOCOL

Staff at the Sandia National Laboratories conducted accelerated aging exposures of the sprinklers from Set 3 in their Facility for Atmospheric Corrosion Testing (FACT II, Figure 1) [6]. In a prior study, observations from metal coupons placed in homes constructed with problem drywall [3] qualitatively matched Battelle Class IV corrosion levels, a corrosion process dominated by sulfide creep [5]. Sprinklers were exposed for four weeks, which roughly represented a field exposure in a light industrial environment of 20 years. The Class IV environment contains H_2S, NO_2, Cl_2, and water vapor. The nominal composition of the inflow was:

- 200 nL/L H_2S
- 200 nL/L NO_2
- 50 nL/L Cl_2
- 75 % relative humidity

The pollutant gasses were supplied using permeation tubes, and mass flow controllers were used to maintain flow. The chamber volume was 300 L, and the total flow was set at 12 L/min. The gas temperature and pressure were nominally 50 °C and 101 kPa, respectively. A schematic of the system is shown in Figure 1. Within the reaction chamber, the frame arms of all sprinklers were oriented in the same direction to ensure that flow characteristics for the gas mixture were consistent.

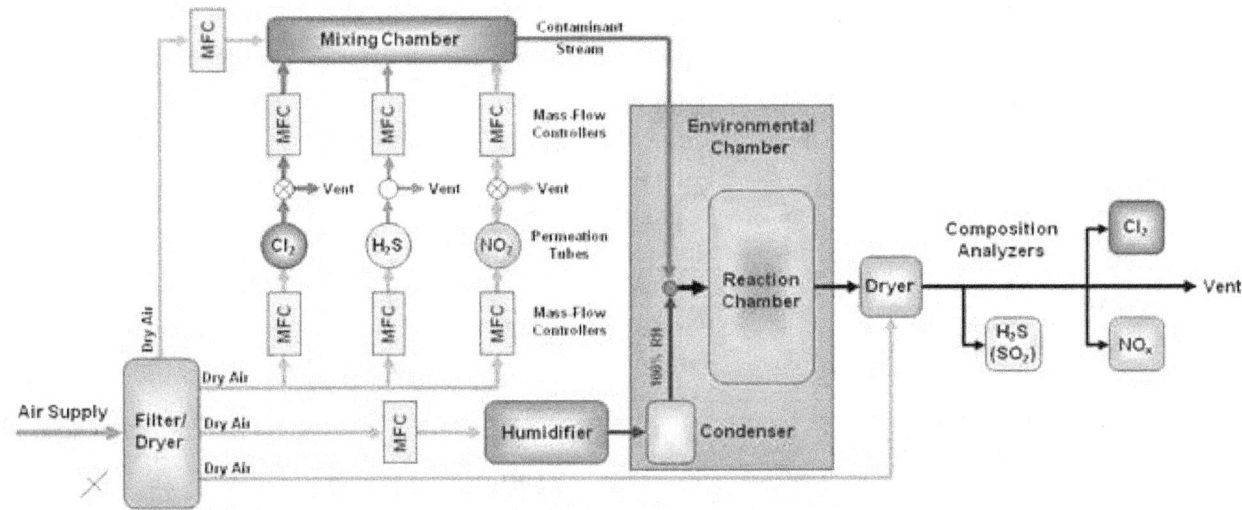

FIGURE 1: FACT II SCHEMATIC

(Figure Courtesy of Sandia National Laboratories, used with permission)

IV. FIRE SPRINKLER SENSITIVITY TEST – PLUNGE TEST

A. PURPOSE

The three major tasks for this research project (*Section III.A*) require the determination of sprinkler activation times. Sprinkler activation time is a key parameter because it defines the point when water is applied to the fire environment, and when the process of fire suppression or control may begin. To obtain the necessary activation time data, NIST tested the sprinklers from Set 1, Set 2, and Set 3 in the UL 199 / UL 1626 plunge test apparatus. NIST staff performed each of the 81 sprinkler activation tests in a plunge test apparatus located at NIST.

The UL 199 standard, which applies to nonresidential sprinklers, and the UL 1626 standard, which applies to residential sprinklers, are sister standards that have many similarities, as well as some differences. Both of the standards, however, contain a sensitivity test that measures the activation time of a sprinkler that is exposed to a stream of heated air. The standards refer to this test as the "oven heat test," which is performed in the "sensitivity test oven." The test is commonly referred to as the plunge test, and is performed in the plunge test apparatus. The plunge test apparatus is a recirculating flow loop device that provides a heated air flow of constant velocity. The subject sprinkler is "plunged" into the hot air flow and the test is terminated when the sprinkler activates. The plunge test apparatus and operating conditions are the same in both standards, and the maximum allowable times for sprinkler activation are consistent between the two for residential and QR sprinklers of the same temperature rating. The bulb-type sprinklers examined in this study were residential sprinklers with a temperature rating of 68 °C (155 °F), which are subject to the UL 1626 standard, and are required to activate in the plunge test in 16.0 s or less. The fusible-type sprinklers in this study were quick response (QR) nonresidential sprinklers with a temperature rating of 74 °C (165 °F), which are subject to the UL 199 standard, and are required to activate in the plunge test in 18.8 s or less.

B. APPARATUS DESCRIPTION

The UL 199 / UL 1626 plunge test apparatus is a flow loop designed to generate a heated air flow across the fire sprinkler, with a constant air velocity of 2.54 m/s ± 0.01 m/s and at a constant temperature of 135 °C ± 1 °C for the subject sprinklers [9, 13]. In the test section, the inside duct axial dimension is approximately 406 mm long, with a rectangular cross section of approximately 203 mm x 203 mm. A variable speed fan is used to generate flow within the duct, and an electrical resistance heater in the heating plenum serves to heat the flow. Fine mesh screens are used on both ends of the measurement section to increase uniformity of the flow over the cross section of the duct. The NIST plunge test apparatus is shown schematically in Figure 2 and photographically in Figure 3. Further details of the plunge test are contained in UL 199 and UL 1626. The modeling and measurement of sprinkler thermal response are summarized in the published literature [14].

The temperature of the air flowing over the sprinkler is measured and controlled using a thermocouple slightly upstream of the measurement section. An orifice plate and pressure taps several duct diameters upstream of the measurement section, in conjunction with a differential pressure transducer, are used to monitor the air velocity within the apparatus. The air velocity at the sprinkler location is measured before testing using a bidirectional probe and differential pressure transducer, while the velocity during testing is monitored via the pressure drop across the orifice plate.

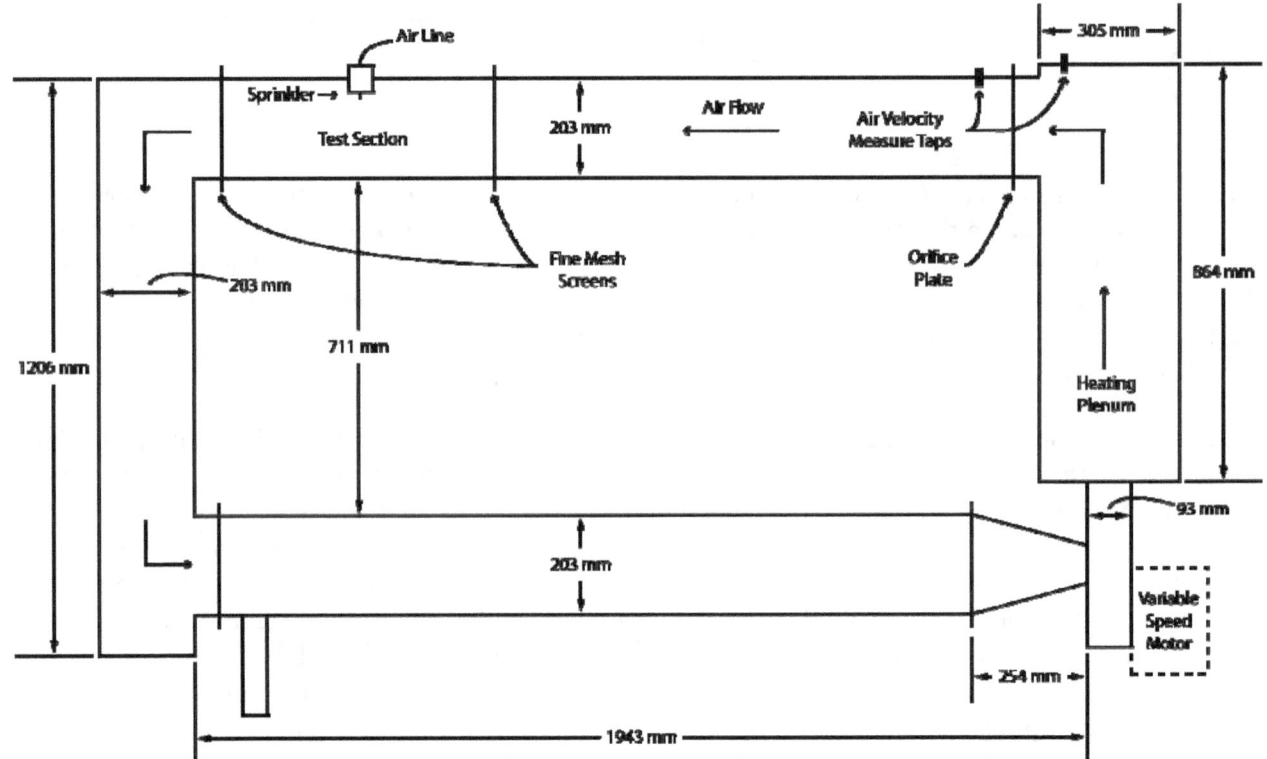

FIGURE 2: SCHEMATIC OF THE NIST PLUNGE TEST APPARATUS[15]

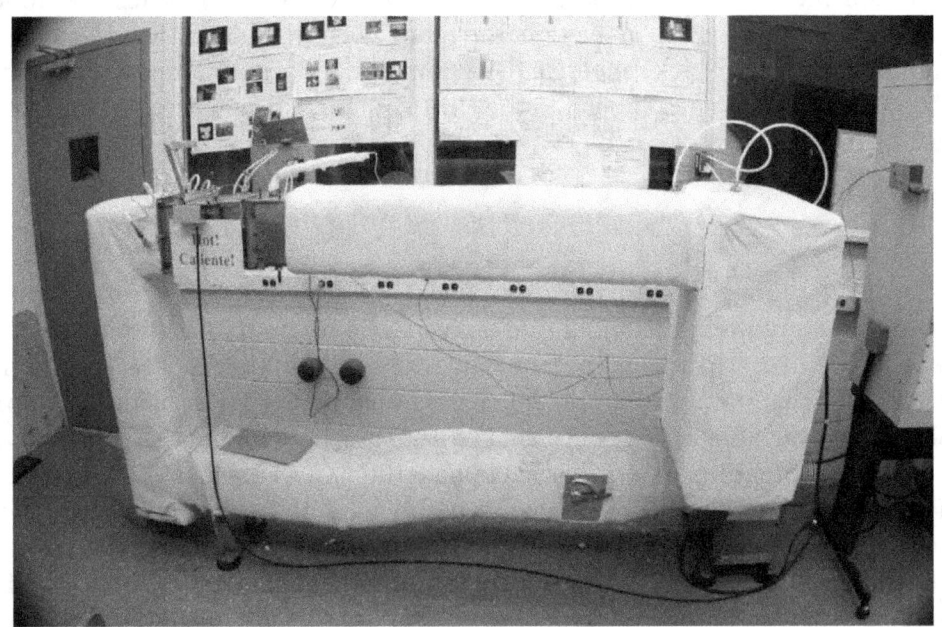

FIGURE 3: PHOTOGRAPH OF THE NIST PLUNGE TEST APPARATUS

The test apparatus includes a variable speed motor controller to provide consistent air velocity. A temperature controller displays the temperature set point, the current temperature of the air flow upstream of the measurement section, and controls the air heating element to provide a constant

10

air temperature. A timing circuit senses the insertion of the sprinkler via a switch on the test section, and monitors air pressure on the sprinkler. A close-up photograph of the test section, showing the thermocouple location and timer start switch, is shown in Figure 4. Figure 5 shows a sprinkler mounted in the sample holder prior to insertion into the heated air flow, while Figure 6 shows the sample holder, with air line installed, during a test. The activation time reported by the plunge test is the elapsed time between the insertion of the sprinkler into the heated air flow and the air pressure drop, which is recorded by the apparatus during testing, due to sprinkler activation. The apparatus pressurizes the sprinkler with air, rather than water, for experimental convenience.

The sprinkler is recorded as "activated" by the apparatus when the air pressure drops from the 55.2 kPa ± 2.8 kPa (8.0 psi ± 0.4 psi) initial gauge pressure to 41.4 kPa ± 2.8 kPa (6.0 psi ± 0.4 psi) gauge pressure, where the expanded uncertainty in the pressure represents a 95 % nominal confidence level, with a coverage factor of 1.96 [16]. The pressure range spans the minimum sprinkler residual gauge pressure of 48 kPa (7 psi) required by NFPA 13 during sprinkler operation. In order to meet the NFPA 13 minimum residual pressure requirement, the initial static gauge pressure at the sprinkler would be greater than 48 kPa (7 psi). This pressure is higher than the sprinkler test gauge air pressure given in UL 199 / UL 1626, which is 28 kPa ±7 kPa (4 psi ±1 psi), but is consistent with the minimum pressure required for sprinkler systems in NFPA 13. Typical static water pressures (gauge) in a wet pipe residential sprinkler system are approximately 276 kPa (40 psi) to 551 kPa (80 psi)[7].

While the activation time is reported to 1 ms by the apparatus, the uncertainty in the activation time may be caused by a variety of factors, acting independently or in concert, including but not limited to: the initial sprinkler temperature, laboratory temperature, air flow temperature, air flow velocity, air pressure drop time, and the time necessary to "plunge" the sprinkler into the air stream. Based on the measurement uncertainties for the oven heat test (plunge test) parameters specified in UL 1626, and assuming a nominal sprinkler temperature rating of 68 °C and a response time index (RTI)[xii] of 50 (m·s)$^{1/2}$, the calculated mean sprinkler activation time is 15.84 s ± 0.28 s, where the expanded uncertainty is characterized by a 95 % nominal confidence level and coverage factor of 1.96. Based on the measurement uncertainties calculated for the NIST plunge test apparatus used in this study, and assuming a nominal sprinkler temperature rating of 68 °C and an RTI of 50 (m·s)$^{1/2}$, the calculated mean sprinkler activation time is 15.86 s ± 0.36 s, where the expanded uncertainty is characterized by a 95 % nominal confidence level and coverage factor of 1.97. Each of these uncertainties is approximately 2 % of the maximum sprinkler activation time specified by UL 1626.

[xii] The sprinkler response time index (RTI) is the product of the thermal time constant of the sprinkler heat actuated element and the square root of the associated gas velocity [14]. Residential and quick response sprinklers are required by NFPA 13 to have a heat actuated element with an RTI of 50 (m·s)$^{1/2}$ or less, where, all other factors being equal, lower values of RTI result in faster sprinkler activation. Sprinkler activation time in the plunge test, t, is calculated by the following equation: $t = -\text{RTI} \cdot (u^{-0.5}) \cdot \ln[1 - [(T_m - T_i) \cdot (T_g - T_i)^{-1}]]$, where T_m is the sprinkler temperature rating, T_g is the temperature of the air stream in the plunge test section, T_i is the initial sprinkler temperature, and u is the nominal gas velocity in the plunge test section [9, 14]. This equation can also be used to calculate the RTI of a sprinkler based on the activation time measured in the plunge test.

FIGURE 4: CLOSE-UP OF THE TEST SECTION

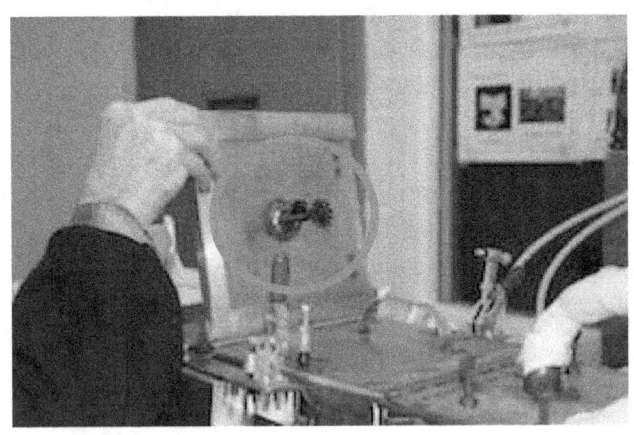

FIGURE 5: SPRINKLER PRIOR TO INSERTION

FIGURE 6: SAMPLE HOLDER WITH AIR LINE
ATTACHED DURING TEST

C. TEST PROTOCOL

An important consideration in designing a test series to discern the presence or absence of changes in properties (here, the activation time of the sprinklers) is the isolation of test factors that might confound that determination. In this study, the following were varied in the testing:

- Test order. The sequence in which individual sprinkler samples of the same type were loaded into the plunge test apparatus was varied, including replicate tests of units of the same model of sprinkler, bulb versus fusible-types, sprinkler finish types, pendent versus sidewall, and new versus aged versus sprinklers recovered from the field. Varying the test order enabled identification of any drift in the test apparatus, separate from other sources of variation, which might have affected test results.

- Day of replicate tests. Replicate tests of sprinklers of the same model and treatment were performed on different days. Varying the day of replicate tests enabled identification of any differences in the test apparatus from day-to-day that might have affected test results. The fusible-type sprinklers were tested on 1/25/2011 and 1/26/2011. The bulb-type sprinklers were tested on 1/28/2011 and 1/31/2011.

The sprinklers were tested according to the oven heat test section of UL 199 / UL 1626. During the 4 days of testing, the mean temperature in the plunge test laboratory was 21.1 °C ± 0.6 °C, where the expanded uncertainty is characterized by a 95 % nominal confidence level and a coverage factor of 1.96. The sprinklers were located in a conditioning room for at least 2 hours prior to each test at 23.5 °C ± 0.6 °C, where the expanded uncertainty is characterized by a 95 % nominal confidence level and a coverage factor of 1.96. The plunge test apparatus was turned on and allowed to heat for 2 hours in order to reach steady state conditions. The air flow in the test section was maintained at a mean velocity of 2.58 m/s ± 0.10 m/s and a mean temperature of 135.0 °C ± 0.6 °C (both with the expanded uncertainty characterized by a 95 % nominal confidence level and a coverage factor of 1.98 over the 4 days of testing). The sample holder with bidirectional probe was inserted into the test section to verify the velocity before testing.

The sprinkler to be tested was installed in the sample holder, without polytetrafluoroethylene (PTFE) tape or thread compound while located in the temperature-controlled sprinkler conditioning room. The sample holder and sprinkler were allowed to equilibrate in the conditioning room for approximately 5 minutes, then transported approximately 50 m in an insulated container to the laboratory, and immediately tested by plunging the sprinkler into the heated air flow. The activation time, air temperature, and air velocity were recorded for later analysis.

In order to avoid the potential impact of sample holder variation on sprinkler activation time, a single sample holder was used for all tests and was allowed to cool in the conditioning room prior to each test.

Details of the test method can be found in UL 199 and UL 1626.

This page intentionally left blank.

V. RESULTS AND ANALYSIS

A. LIMITATIONS

The findings of this report are subject to certain limitations that should be considered when interpreting the data.

Only nine Set 1 residential sprinklers (provided by CPSC) were available for analysis. The findings cannot be generalized to the universe of all sprinklers exposed to the emissions of problem drywall because no formal statistical sampling methods were used, and the sample size was small relative to the number of exposed sprinklers. The sprinklers collected by the CPSC from homes with problem drywall consisted of only bulb-type sprinklers from one manufacturer; therefore, conclusions about the effects of in-home contamination on other manufacturers and types of sprinklers are not possible.

The Set 1 sprinklers were subject to unknown environments for approximately three years prior to testing at NIST. The indoor contaminant concentrations to which a particular sprinkler may be exposed is a function of such factors as the nature and frequency of household cooking; indoor production of aerosols/particles/vapors from items such as paint, air fresheners, household chemicals or pets; frequency of air changes within the home; local outdoor air quality; and the occurrence of smoking within the home. In addition, the specifics of the plumbing network and sprinkler water supply, the sprinkler installation and removal processes, and sprinkler handling procedures for the Set 1 sprinklers have not been studied. These factors are independent of the nature of the emissions from the drywall and might have affected the sprinklers' performance.

The applicability of the Battelle Class IV corrosion accelerated aging protocol performed at Sandia National Laboratories is not evaluated here. However, the conditions listed in the prior paragraph are not part of the Battelle Class IV protocol, so the relationship between the Set 1/Set 2 sprinklers and the relationship between the Set 2/Set 3 sprinklers are not the same. Any such comparison is confounded further by differences between the models of sprinklers available for testing. The plunge tests conducted for new sprinklers and sprinklers that underwent accelerated aging (Set 2 and Set 3, respectively) examined only the isolated effect of the incremental chemical contaminants expected from the problem drywall over a period designed to simulate 20 years of exposure.

There are experimental conditions that may affect sprinkler activation time. These include, but are not limited to, the construction of the sample holder and the use of PTFE tape or pipe thread sealant on the sprinkler threads. While these variables have been controlled in this study, the absolute activation time results may differ among test laboratories due to these and other factors. For this reason, the relative results of the study are more useful than the absolute activation times.

The results cannot be extrapolated to all types of fires. There are many characteristics of fires that may affect sprinkler performance, including, but not limited to the fire growth rate, fire location, sprinkler location, room geometry, and water supply.

NIST did not assess the change in hazard to occupants or firefighters resulting from changes in sprinkler performance.

1. SPRINKLER ACTIVATION TIME

To address the first goal of this study (Do sprinklers exposed to problem drywall emissions activate within allowable time specifications?), the absolute sprinkler activation times of the nine sprinklers provided by CPSC (Set 1) were measured and assessed. All of these sprinklers were bulb-type sprinklers of a single model from Manufacturer 3. The results of the nine tests[xiii] are shown graphically in Figure 7 and demonstrate that all the sprinklers from Set 1 activated in the plunge test within the maximum allowable activation time specified in the UL 1626 standard. (The data for sprinklers from Set 1 are the circular points labeled "Field.") The gray bar represents the mean of the data, and the horizontal black bar represents the maximum allowable activation time of 16.0 s for residential sprinklers of their temperature rating, as specified by UL 1626.

The absolute sprinkler activation times for the new sprinklers (Set 2) and the sprinklers subjected to accelerated aging (Set 3) are also plotted in Figure 7. The mean activation time for each set of data is indicated by the associated gray bar. The figure illustrates that all of the bulb-type sprinklers activated at times less than 16.0 s.

Of the fusible-type sprinklers tested, all but one activated at times less than 18.8 s, the maximum plunge test activation time specified by UL 199 for nonresidential QR sprinklers of their temperature rating. One of the aged fusible sprinklers from Manufacturer 3 did not activate successfully during the plunge test, and the test was terminated at 600 s. The result from the nonactivating sprinkler was treated as a missing data point and excluded from all further analyses[xiv].

Statistical analysis is useful for examining the information gathered from the measurements of the sample sprinkler activation times. The UL 199 / UL 1626 test methods, for example, specify the plunge test sample sizes for new sprinklers as 10 units, and the test sample size for sprinklers subjected to corrosive exposures as 5 units. The overall goal is to provide confidence that a very high percentage of the sprinklers installed in buildings will meet the specified maximum activation time criterion.

Figure 8 shows the statistical tolerance limits calculated from the test data for the sprinklers tested. The upper tolerance limit represents the upper bound of activation time that includes 99 % of the activation times for the population of sprinklers, with 95 % confidence. In other words, it is expected that 99 % of the sprinkler population, which includes all sprinklers of a particular model and treatment, manufactured under the same conditions, past, present, and future, would have plunge test activation times equal to or less than the upper tolerance limit. Further details of the tolerance limit calculations are included in Appendix A and Table D.1 in Appendix D.

[xiii] Some of the activation time values are very close or identical, making it appear that there are fewer triangles, asterisks, or circles than actually measured in each set of data for a certain type of sprinkler made by a particular manufacturer.
[xiv] The result from the non-activating sprinkler was not treated as a censored observation because the reason for its non-activation is unknown and could have a cause, such as a manufacturing defect, with a failure mechanism that is not relevant to this study.

Figure 8 illustrates that for the seven types of sprinklers, including the sprinklers provided by the CPSC, the activation time upper tolerance limits are near the maximum activation times specified in the UL standards. If additional data were available, however, the bounds would be tighter, on average, because the coverage factor for these tolerance bounds [18] is a decreasing function of the number of measurements. Thus, while these results do not show that all different types of sprinklers meet the maximum activation time specification, they do not necessarily rule out that possibility. Further details on the analysis of the absolute sprinkler activation times, descriptive statistics, and tolerance limits are contained in Appendix A through Appendix D.

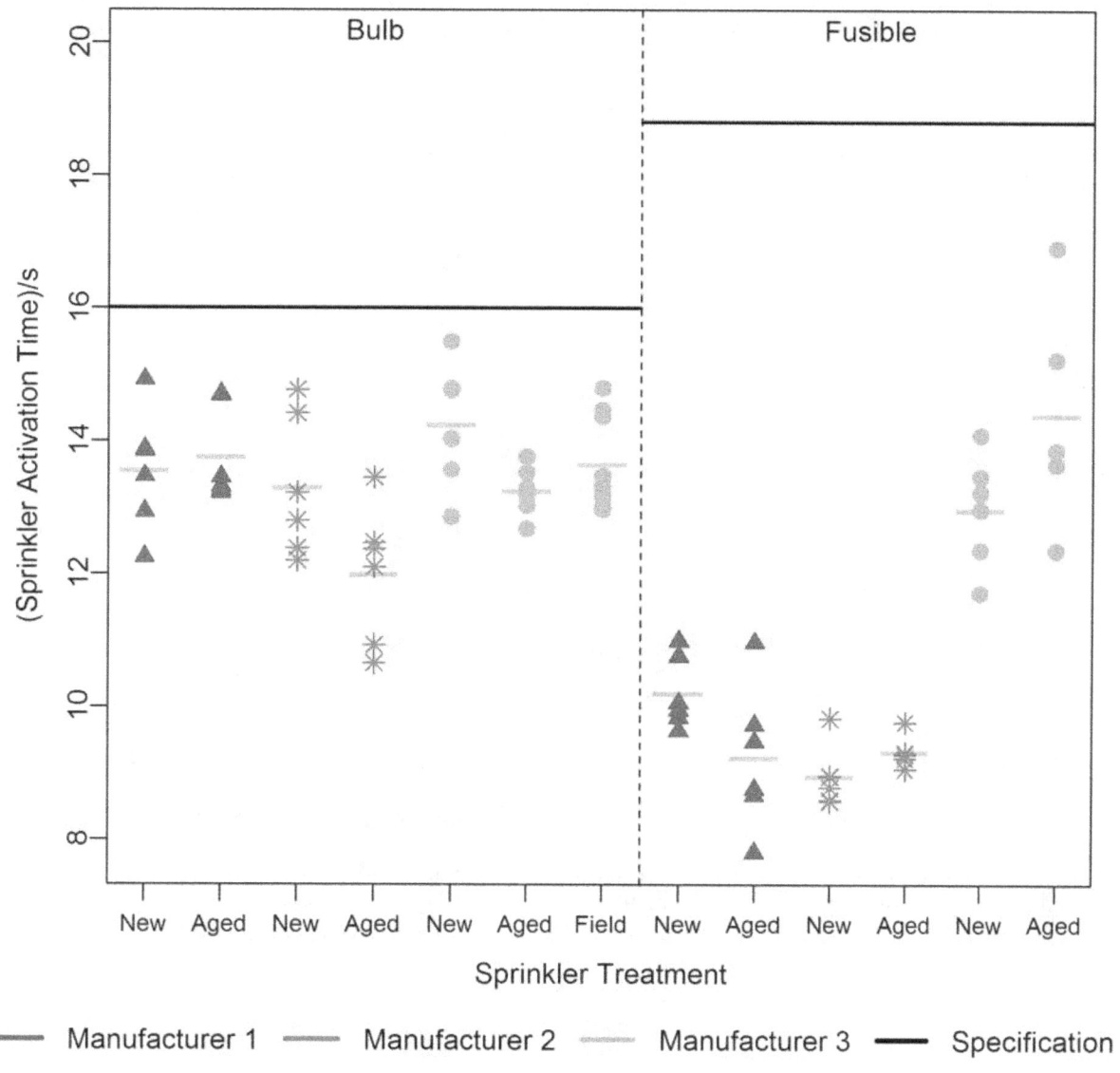

FIGURE 7: RESULTS OF UL 199 / UL 1626 PLUNGE TESTS.

Sprinkler activation times are plotted versus sprinkler treatment. The left-hand section of the plot shows the results for bulb-type sprinklers, while the right-hand section shows the results for fusible-type sprinklers. The black reference lines at 16.0 s and 18.8 s are the performance specifications for the activation times for each type of sprinkler, as specified by UL 1626 and UL 199, respectively. The gray line in the middle of each set of observed activation times is the mean activation time for that set of data.

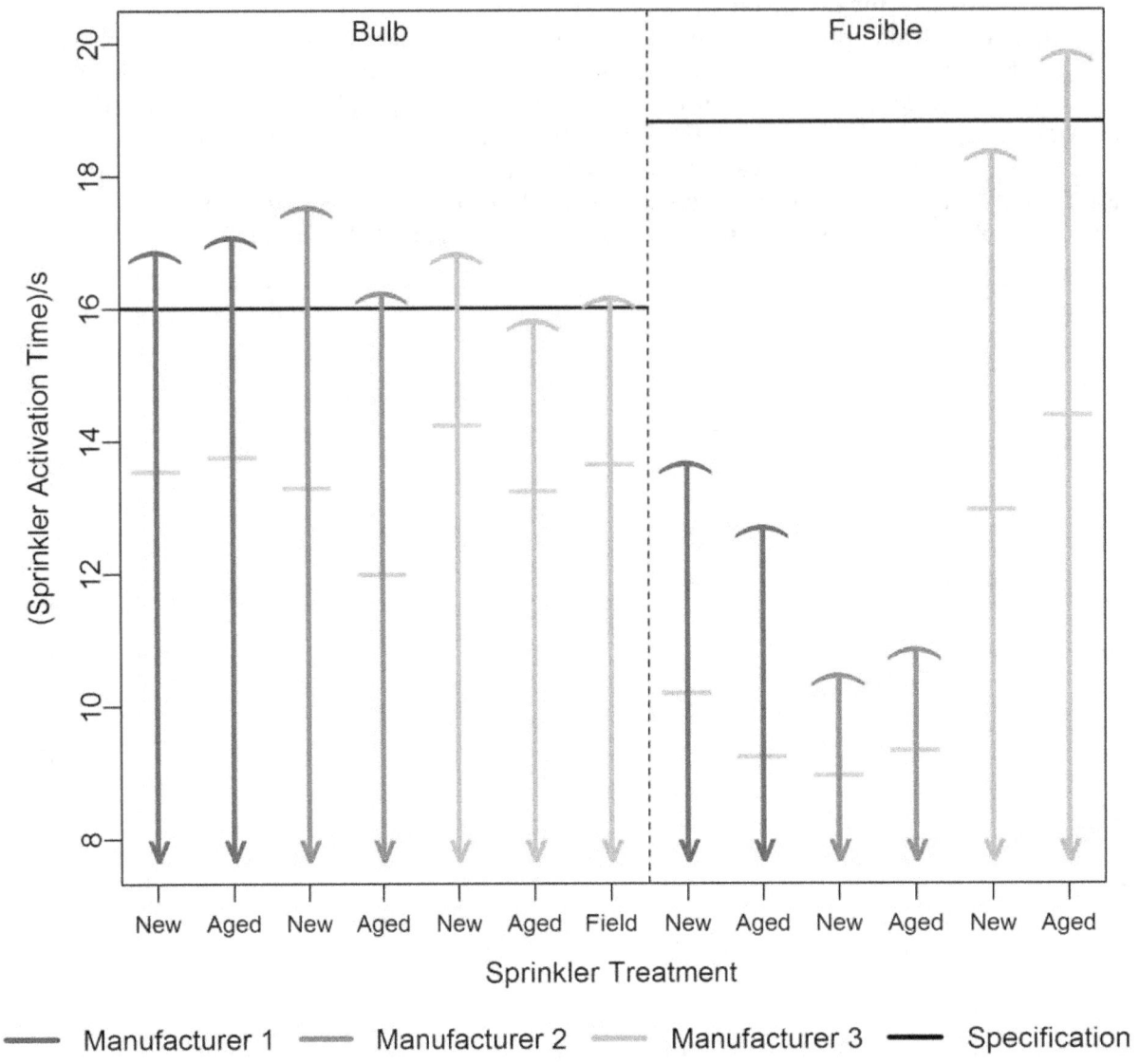

FIGURE 8: STATISTICAL TOLERANCE LIMITS FOR SPRINKLER ACTIVATION TIME.

The curved upper bound shown for each population of sprinkler activation times will exceed the activation times of 99 % of all sprinklers from the associated population with 95 % confidence. The gray line associated with each tolerance limit is the sample mean of the observed activation times for the associated set of data.[xv]

[xv] Note that the tolerance limits for the bulb-type sprinklers from Manufacturer 3 shown here were computed without taking into account the small, but statistically significant, day effect found for these sprinklers. However, additional analyses carried out to study the effect of this simplification show that there is no practical difference in the results whether not the day effect is included. . For details, see Appendix A, Table D.1, Figure D.2, and Tables E.3 and E.4.

2. COMPARISON OF RECOVERED AND NEW SPRINKLERS

In order to address the study's second goal (Do sprinklers exposed to problem drywall emissions activate differently from new sprinklers of the same model?), the UL 1626 plunge test activation time results for the sprinklers in Set 1 were compared with new sprinklers of the same manufacturer model. Note from Table 1 that the sprinklers provided by the CPSC and the new sprinklers were all from Manufacturer 3 with the same model number, but were manufactured in different years and had sprinkler bodies with different finishes.

The summary of the activation time data is shown in Table 2. The mean activation time for the bulb-type sprinklers provided by the CPSC is 13.6 s, while the mean activation time for the same model of new sprinklers is 14.2 s. The maximum activation time specified in UL 1626 for these sprinklers is 16.0 s. While the difference in the mean activation times of the new and the field sprinklers is approximately 4 %, the analysis of the data indicates that there is no statistical difference between the activation times of the new sprinklers (Set 2) and those provided by the CPSC (Set 1) at the 95 % confidence level (an individual t-test of this hypothesis has a p-value of $p = 0.22$). Although we call out the specific comparison that addresses the second goal of the study here, potential effects associated with all three sets of sprinkler exposures were addressed in the statistical analysis, details of which are included in Appendix A, Table E.3, and Table E.4.

Bulb-type Manufacturer	Set 2 (new) (Mean/s)	Set 2 (new) (St Dev/s)	Set 1 (field) (Mean/s)	Set 1 (field) (St Dev/s)	Observed Difference (s)	Observed Difference (%)
3	14.2	1.0	13.6	0.7	0.6	4.2 %

TABLE 2: SUMMARY OF PLUNGE TEST DATA FOR SET 1 AND SET 2 SPRINKLERS OF THE SAME MODEL

3. COMPARISON OF ACTIVATION TIMES OF NEW AND AGED SPRINKLERS

A. BULB-TYPE SPRINKLERS

To address the third goal of this study (Does a 20-year exposure to a Battelle Class IV corrosivity environment, simulated through an accelerated aging test protocol, affect sprinkler activation time performance?), the activation time results for the bulb-type sprinklers in Set 2 (new) and Set 3 (aged), subjected to the UL 1626 plunge test, are shown in Figure 7. There were six plunge tests for each new and aged treatment, but some activation times were very similar, resulting in multiple data points on top of each other. Gray lines in Figure 7 indicate the mean value of the activation time, while the black line indicates the maximum allowable activation time of 16.0 s for 68 °C (155 °F) rated residential sprinklers per UL 1626. (Note that the testing of these types of sprinklers is destructive, so the same sprinkler cannot be tested as new, and then again as aged.)

A summary of the plunge test data for bulb-type sprinklers is shown in Table 3, comparing the activation of new sprinklers to sprinklers subjected to accelerated aging. The analysis indicates that there is an observed difference of up to approximately 10 % between the sprinkler treatments. (Note that a negative difference indicates that the aged sprinkler responds more slowly than the

19

new sprinkler.) Further analysis, however, revealed that at the 95 % confidence level, the only statistically significant difference was between the activation times for the new and aged sprinklers from Manufacturer 3 (see Table E.3 in Appendix E for the p-value associated with the statistical test for the effect of this factor). For bulb-type sprinklers from Manufacturer 3, the new sprinklers were found to activate approximately 1 s slower than the aged sprinklers, with the true difference likely between 0.22 s and 1.79 s (95 % confidence level). Details of the statistical analysis are included in Appendix A, Table E.3, and Table E.4.

Analysis also indicated that at the 95 % confidence level, there is a statistically significant difference between the activation times for the bulb-type sprinklers from Manufacturer 3 on the first and second days of testing (see Table E.3 in Appendix E for the p-value associated with the statistical test for the effect of this factor). Sprinkler activation times resulting from tests performed on 1/31/2011 were approximately 0.91 s longer, on average, than activation times from tests performed on 1/28/2011, with the true difference likely between 0.42 s and 1.40 s (95 % confidence interval). This effect was not seen for the other two sprinkler manufacturers, and was also not seen for the fusible-type sprinklers (*Section V.B.3.B*). This difference did not detrimentally affect the results due to the design of the study and experimental test sequence, which used the day as a blocking factor. Therefore, the effect of sprinkler treatment on activation time could be studied using data from each test day individually, with the results averaged over two days.   Supporting statistical detail is provided in Appendices A, B, and E.

Mfr Bulb Type	Set 2 (new) (Mean/s)	Set 2 (new) (St Dev/s)	Set 3 (aged) (Mean/s)	Set 3 (aged) (St Dev/s)	Observed Difference (s)	Observed Difference (%)
1	13.5	0.9	13.8	0.7	-0.2	-1.6 %
2	13.3	1.1	12.0	1.0	1.3	9.8 %
3	14.2	1.0	13.2	0.4	1.0	7.1 %

TABLE 3: SUMMARY OF PLUNGE TEST DATA FOR SET 2 AND SET 3 BULB-TYPE SPRINKLERS

B. FUSIBLE-TYPE SPRINKLERS

The activation time results for the fusible-type sprinklers in Set 2 (new) and Set 3 (aged), subjected to the UL 199 plunge test, are shown in Figure 7. There were 6 plunge tests for each new and aged treatment; however, some activation times were very similar, resulting in multiple data points near or on top of each other. In addition, one of the sprinklers from Manufacturer 3, subjected to the accelerated aging protocol, did not activate during testing, resulting in one less data point. Gray lines in Figure 7 indicate the mean value of the activation time, while the black line indicates the maximum allowable activation time of 18.8 s for 74 °C (165 °F) rated QR sprinklers per UL 199. The testing of these types of sprinklers is destructive, so the same sprinkler cannot be tested as new and then again as aged.

A summary of the plunge test data is shown in Table 4 for the fusible-type sprinklers, showing the activation time of new sprinklers compared to sprinklers subjected to accelerated aging. The analysis indicates that there is an observed difference of up to approximately 11 % between the new and the aged sprinklers. For Manufacturer 1, the aged sprinklers activated earlier than the new sprinklers. For Manufacturer 2 and Manufacturer 3, the aged sprinklers activated later than the new sprinklers. Further analysis, however, revealed that at the 95 % confidence level, there is not a statistically significant difference between the activation times of new and aged sprinklers for any manufacturer. Supporting statistical detail is provided in Appendices A, C, and F.

As discussed in SectionV.B.3.A above, the analysis indicated that at the 95 % confidence level, there was not a statistically significant difference between the activation times for the fusible-type sprinklers on the first and second days of testing. Supporting statistical detail is provided in Appendices A, C, and F.

Mfr Fusible Type	Set 2 (new) (Mean/s)	Set 2 (new) (St Dev/s)	Set 3 (aged) (Mean/s)	Set 3 (aged) (St Dev/s)	Observed Difference (s)	Observed Difference (%)
1	10.2	0.5	9.2	1.1	1.0	9.5%
2	9.0	0.5	9.3	0.2	-0.4	-4.2%
3	13.0	0.8	14.4	1.7	-1.4	-11.0%

TABLE 4: SUMMARY OF PLUNGE TEST DATA FOR SET 2 AND SET 3 FUSIBLE-TYPE SPRINKLERS

This page intentionally left blank.

VI. CONCLUSIONS

This analysis provided answers to three questions:

1. Do sprinklers exposed to the emissions from problem drywall activate within allowable time specifications?

   With the important limitations listed in Section V.A, all residential sprinklers in Set 1 (bulb-type from the field), tested in the plunge test apparatus activated within the maximum permissible activation time limit of UL 1626. Given the sprinklers available from the field provided for testing, NIST was not able to conclude whether these sprinklers would continue to activate within intended specifications if exposed to problem drywall emissions for the 20-year testing or replacement interval (5 years for corrosive environments) required by NFPA 25 for sprinklers with fast response elements [17].

   No fusible-type sprinklers from homes were available for testing.

2. Do sprinklers exposed to emissions from problem drywall activate differently than new sprinklers of the same manufacturer and model?

   With the important limitations listed in Section V.A, NIST did not find a statistically significant difference between the activation times of sprinklers in Set 1 (bulb-type from the field) and Set 2 (new bulb-type sprinklers) of the same manufacturer and model.

   No fusible-type sprinklers from homes were available for testing.

3. Does a 20-year exposure to a Battelle Class IV corrosivity environment, simulated through an accelerated aging test protocol, affect sprinkler activation time performance?

   With the important limitations listed in Section V.A, NIST measured a small, but statistically significant decrease in the time required to activate the tested bulb-type sprinklers from Manufacturer 3 as a result of a 4-week exposure to a Battelle Class IV corrosivity environment designed to simulate a 20-year exposure to the emissions from problem drywall. The approximate 1 s change in activation time is small compared to the maximum specified activation time of 16.0 s.

   With the important limitations listed in Section V.A, NIST did not find a statistically significant difference in activation time of new bulb-type sprinklers from Manufacturer 1 and Manufacturer 2 versus the same model sprinklers that were aged by a 4-week exposure to a Battelle Class IV corrosivity environment designed to simulate a 20-year exposure to the emissions from problem drywall.

   With the important limitations listed in Section V.A, NIST did not find a statistically significant difference in activation time of new fusible-type sprinklers from any of the three manufacturers tested versus the same model sprinklers that were subjected to the simulated aging protocol. This comparison did not include the one Set 3 sprinkler from Manufacturer 3 that did not activate.

   With the important limitations listed in Section V.A, NIST found that a single fusible-type sprinkler from Set 3, Manufacturer 3, did not activate after 600 s of exposure in the plunge test apparatus. Because each sprinkler could be tested only once, it could not be

23

determined whether the failure to activate was inherent in that particular unit or whether the failure to activate was due to the accelerated aging exposure in the Battelle Class IV corrosivity environment.

VII. ACKNOWLEDGMENTS

The authors gratefully acknowledge the contributions of several individuals. The experience of Jay McElroy (NIST) in running the plunge test apparatus was instrumental in the success of the project. Dr. Rob Sorenson (Sandia National Laboratories) led the accelerated aging exposures (Set 3 sprinklers) and provided (through CPSC) the description of the accelerated aging method contained in this report.

This page intentionally left blank.

APPENDIX A: OVERVIEW OF STATISTICAL METHODS AND DATA ANALYSIS

1. EXPLORATORY DATA ANALYSIS

For each set of sprinkler activation data for sprinklers sharing a common operating principle, NIST first performed a graphical exploratory data analysis to see what factors influenced the responses and to identify any unusual data points or other interesting features in the data. The exploratory analysis informs all subsequent statistical analyses and may influence the quantitative analyses that had been planned prior to the collection of the data, if the assumptions underlying the planned analyses do not match the characteristics of the data obtained.

The factors examined in this stage included test date, run order, uniformity of activation times by manufacturer's model, and the treatment of each sprinkler, new, aged, or taken from the field. The output from the exploratory examination of the sprinkler activation time data for bulb- and fusible-type sprinklers are presented in Appendices B and C, respectively. Specific findings or conclusions about the results from this analysis are given in the captions accompanying each figure in these appendices. Comparing the activation times of sprinklers subject to different treatments across types and manufacturers as shown in Appendices B and C suggests that there is no single underlying effect of aging on sprinkler activation time that is being overlooked in each individual set of data because of the limited sample sizes at that level. This conclusion is based on the fact that the new sprinklers sometimes appear to activate faster than the aged sprinklers and sometimes appear to activate more slowly.

2. QUANTITATIVE ANALYSIS

The second phase of the statistical analysis is a numerical quantification of absolute sprinkler performance, as defined by sprinkler activation times and quantification of differences in activation times between sprinklers subject to different treatments. Analyses of absolute sprinkler response times for both types of sprinklers are given in Appendix D. The analysis of different factor effects on sprinkler response times is given in Appendices E and F for bulb- and fusible-type sprinklers, respectively.

To quantify the absolute sprinkler performance, NIST computed upper statistical tolerance limits [Reference 18, eq. 1.3 and eq. 1.4] for sprinkler activation times by sprinkler operating principle, manufacturer, and treatment. These limits bound from above or below a desired proportion of all future values from a population with a specified level of confidence. In this case, the populations of interest are the populations of measured sprinkler activation times for sprinklers that operate via a bulb or fusible mechanism, were manufactured by one of the companies whose sprinklers were tested in this study, and were either new, subjected to an accelerated aging procedure at Sandia National Laboratories, or taken from the field. Upper tolerance limits were computed for 13 different populations in all[xvi]. The resulting tolerance limits should each exceed at least 99 % of all

---

[xvi] All of the tolerance intervals shown in Table D.1 and Figure D.2 are based on the averages of data collected over two days of testing. As determined in the comparison of the factor effects, small, but statistically significant, day-to-day variations were observed in the activation times for the bulb sprinklers made by Manufacturer 3. Since the measurements

future measured activation times within each population with 95 % confidence. These results are shown in Appendix D.

To quantify potential differences in sprinkler activation times depending on the treatment of the sprinkler, NIST used two types of statistical methods. First, a linear model with qualitative factors was fit to the sprinkler activation data from each manufacturer for each operating principle. Each model included the factors test date and treatment and allowed statistical tests to be made for the significance of any differences in sprinkler activation time observed between days or between treatments (*i.e.*, sets of sprinklers subject to one of the three different environmental exposures).

Although other sprinkler properties, such as operating principle, manufacturer, and finish were not included as factors in the model, the fact that they were used to define the various populations within which day and treatment effects were assessed, allows the scope of the overall study to address these issues through the parallel analyses carried out for each set of data. Fitting the model to the data from each manufacturer and for each operating principle separately has the advantage of not requiring the assumption of equal levels of random measurement error across all sprinklers. The models were fit using analysis of variance [19, 22], or "ANOVA." These analyses are given in Appendices E and F for bulb- and fusible-type sprinklers, respectively.

Subsequently, for those data sets that showed evidence of statistically significant differences between the activation times of sprinklers subject to different levels of each factor, simultaneous confidence intervals for the pairwise differences in the mean responses associated with the levels of each factor were computed using Tukey's "Honest Significant Difference" procedure [19, 20, 21, 22].

In the fit of each model, the factor test date was used as a blocking factor within which comparisons of the different treatments can be made with reduced random variation. The use of test date as a blocking factor requires the assumption of no interaction between the treatments and test date. This assumption seems reasonable from a physical perspective because the presence of an interaction would mean that the treatment effect for a given manufacturer and type of sprinkler would vary from day to day, which does not seem likely to be the case. Graphical analysis of the residuals from the model fit to each set of data, discussed more below, further confirms that a model with no interaction does appear to fit the data well in each case.

In mathematical terms, the model fit to each set of data is specified as:

$$T_{ijk} = \mu + \Delta_i + \tau_j + \varepsilon_{ijk}$$

where T_{ijk} is the measured activation time on test date i for sprinkler k from treatment j,

μ is the mean activation time,

Δ_i is the test date effect for test date i,

---

were made over two days only, however, the data does not have much information that can be used to empirically assess the uncertainty in the results from this source and no day-to-day variance component has been included in these computations. To be sure this simplification has not skewed the results, tolerance bounds were also computing for bulb-type sprinklers from Manufacturer 3 using only the data from 1/31/2011, which had slower response times on average. These results are reported in the footnote to Table D.1. Note that this day effect does not affect the analysis used to compare how the activation times change with different treatment levels, where day-to-day differences are treated as a blocking factor for which any potential effect cancels out.

τ_j is the sprinkler treatment effect for treatment j, and

ε_{ijk} is the random error in the measured activation time for sprinkler k subject to treatment j on test date i.

The random measurement errors, ε_{ijk}, are assumed to follow a normal distribution with a true mean value of zero and a true standard deviation of σ.

Prior to interpreting the statistical tests based on the fit of the model, a graphical residual analysis was carried out to ensure that the assumptions underlying the analysis hold, at least approximately, for the data that has been observed. Appropriateness of the assumptions is indicated when the residuals (1) appear to be randomly distributed with a mean of zero and constant variance across all treatments and test dates, and (2) appear to follow a normal distribution. These plots are shown in Appendices E and F for bulb-type and fusible-type sprinklers, respectively, with commentary on the interpretation of the results given in the caption for each plot. Note that when assessing the constancy of the random variation in the residuals, the values are subject to a fair amount of fluctuation due to random measurement variation. For sample sizes similar to those used in this study, the ratio of two residual standard deviations would have to be smaller than about 1/3, or larger than about 3, to be statistically distinguishable.

The statistical tests for the significance of day-to-day and treatment-to-treatment effects are also presented in Appendices E and F for bulb- and fusible-type sprinklers, respectively. The tests are presented in ANOVA tables that summarize the numerical output from the fit of the model as well. Each ANOVA table essentially includes three variances, one associated with each of the two factors, test date and treatment, and one associated with the random measurement variation in the activation times. These variances, labeled "Mean Squares" in the output from the computations, each capture the variation in the data that arises from random measurement error, plus the factor associated with each line in the ANOVA table.

Each variance or mean square is computed as a sum of squared deviations of different summaries of the measurement data from their associated mean values, divided by its degrees of freedom. The number of degrees of freedom associated with each mean square is essentially an effective sample size of independent deviations for each mean square. When none of the factors have significant effects, all three mean squares will take on the same value on average, and their ratios will follow the F probability distribution. The column labeled "F Value" in the output shows the values of the test statistics, each computed by taking the ratio of the mean square associated with a particular factor to the mean square for residuals alone. If the test statistic value is near 1, it indicates that there is no evidence that the factor being tested (either test date or treatment) adds significant variation to the variation expected from random measurement error alone. On the other hand, if the value of any test statistic is much larger than 1, a statistically significant effect is indicated for the factor in question.

The formal hypotheses being compared by these statistical tests are as follows. For the factor test date the null hypothesis is H_0: $\Delta_1 = \Delta_2 = 0$ versus the alternative hypothesis H_A: $\Delta_i \neq 0$ for some i, $i = 1, 2$. For the factor treatment, which has the levels "New", "Aged", and "Field" for some manufacturers of sprinklers and just "New" and "Aged" for other manufacturers, the hypotheses being compared are either H_0: $\tau_1 = \tau_2 = \tau_3 = 0$ versus H_A: $\Delta_i \neq 0$ for some i, $i = 1, 2, 3$ or

29

H_0: $\tau_1 = \tau_2 = 0$ versus H_A: $\Delta_i \neq 0$ for some i, $i = 1, 2$, depending on the number of levels present in each set of data. Thus, due to the nature of the hypotheses being compared, the statistical tests given in each ANOVA table only tell us which factors have one or more non-zero effects but do not tell us specifically how the responses associated with different levels of each factor compare. (That information is determined in the next phase of the analysis, however, and is described two paragraphs below.)

The determination of how large the test statistic needs to be for an effect to be declared statistically significant is indicated by the values in the last column of each table labeled "Pr(>F)," which gives the p-value of the test. The p-value is the probability of seeing the observed test result, or a more extreme result, by chance. When the p-value is below a predetermined probabilistic threshold termed the significance level, it indicates that the effect is statistically significant at that level. The symbol α is generally used to denote the significance level, and the most commonly used value is $\alpha = 0.05$.

Specific findings or conclusions about the significance of the factors for each set of data are given in the captions for each ANOVA table in Appendices E and F.

After carrying out statistical tests for the significance of the different factors under study, additional follow-up analyses were done for the factors in each set of data that exhibited differences at the $\alpha = 0.05$ significance level. As mentioned above, this analysis allows us to compare how the responses associated with different factor levels are related.

The comparison is carried out by computing simultaneous confidence intervals for pairwise differences of the mean responses for different factor levels. Confidence intervals that do not contain the value zero indicate significant differences between the activation times associated with different treatments. These confidence intervals were computed using Tukey's "Honest Significant Difference" [20, 21], and each set of pairwise intervals simultaneously captures all true pairwise differences with 95 % confidence. These confidence intervals are computed using the differences between all pairs of observed treatment means, the pooled standard deviations for each set of treatment data, the sample sizes associated with each treatment mean, and a coverage factor obtained from the studentized range distribution [21, 22]. The pairwise confidence intervals for each data set with significant treatment differences are displayed with their associated ANOVA tables in Appendices E and F.

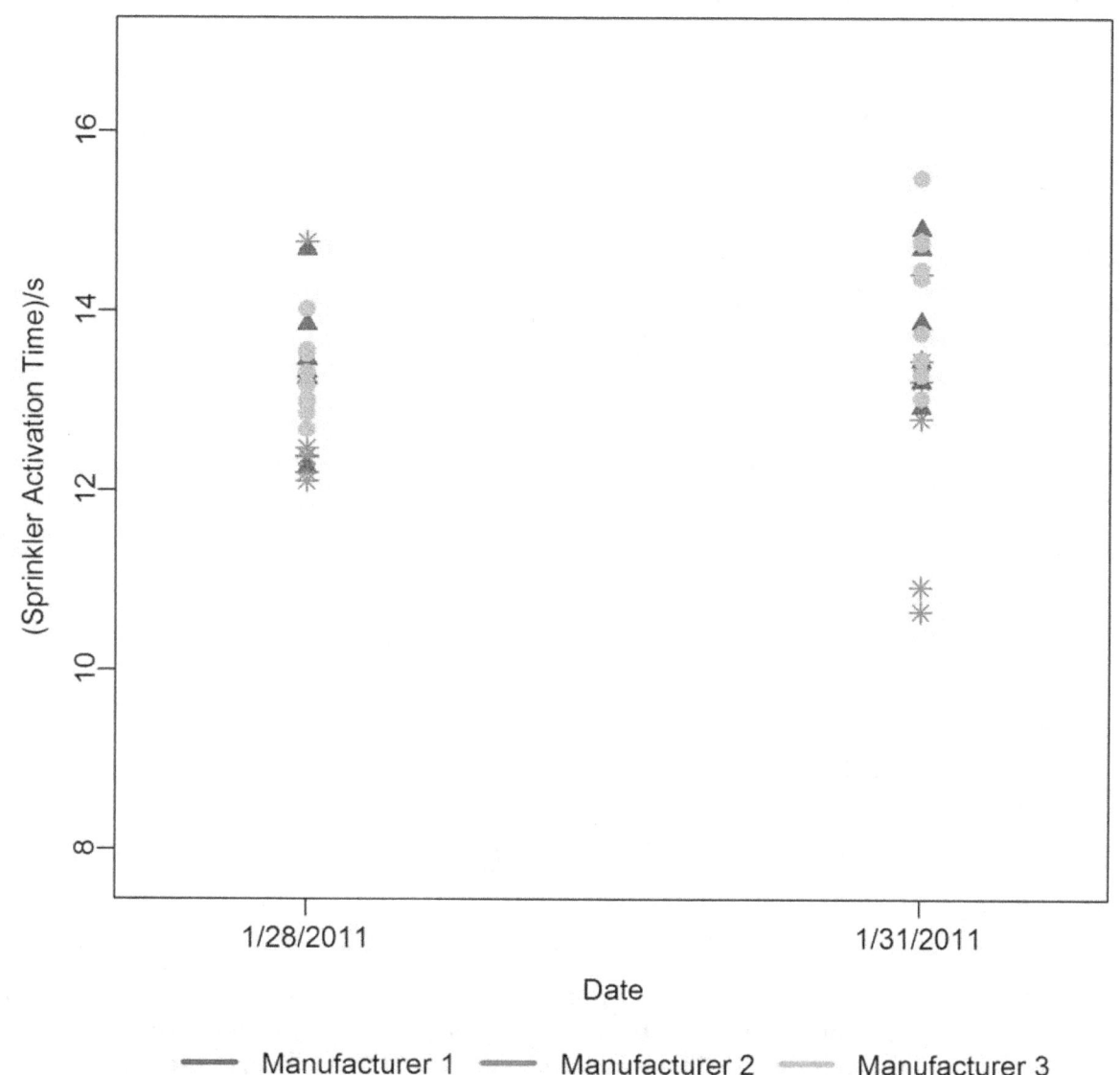

Figure B.1: Sprinkler activation times plotted versus test date for bulb-type sprinklers (Sets 1, 2, and 3). Sprinkler manufacturers are indicated by color coding. This plot shows that the activation times are fairly similar on both test dates, but suggests the results obtained on 1/31/2011, may be a little higher on average than the results obtained on 1/28/2011. In addition, the plot identifies two sprinklers made by Manufacturer 2 that may have activation times that are unusually low. However, Figure B.4 and further analysis shown in Appendix D, ultimately suggest that the results for these two sprinklers are not inconsistent with the results for other sprinklers from the same data set.

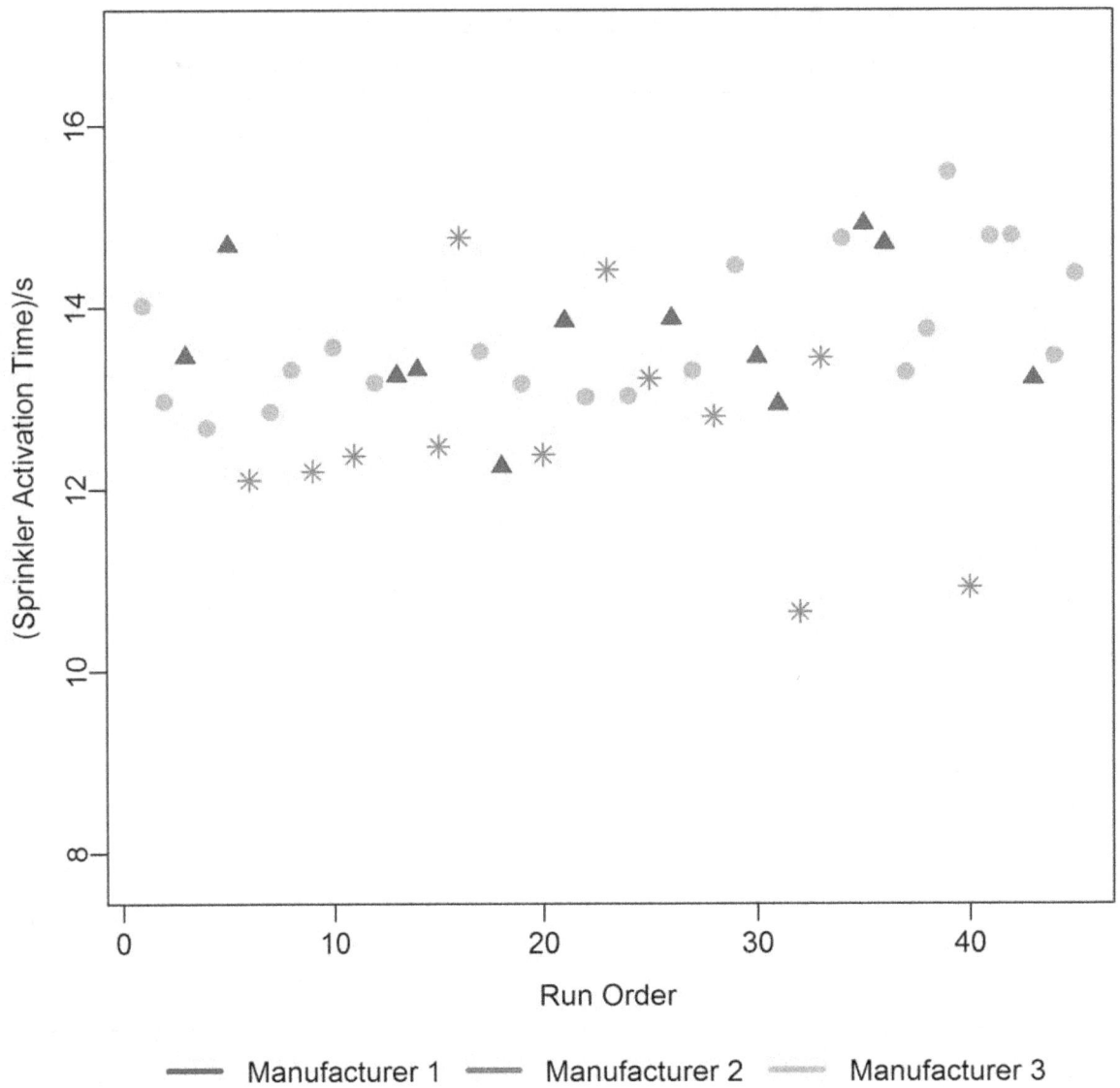

Figure B.2: Sprinkler activation times versus run order for bulb-type sprinklers (Sets 1, 2, and 3). No drift over time is apparent in this plot, although the slight increase in response times seen in Figure B.1 is visible in this plot as well. The two sprinklers from Manufacturer 2 with the lowest activation times also stand out in this plot.

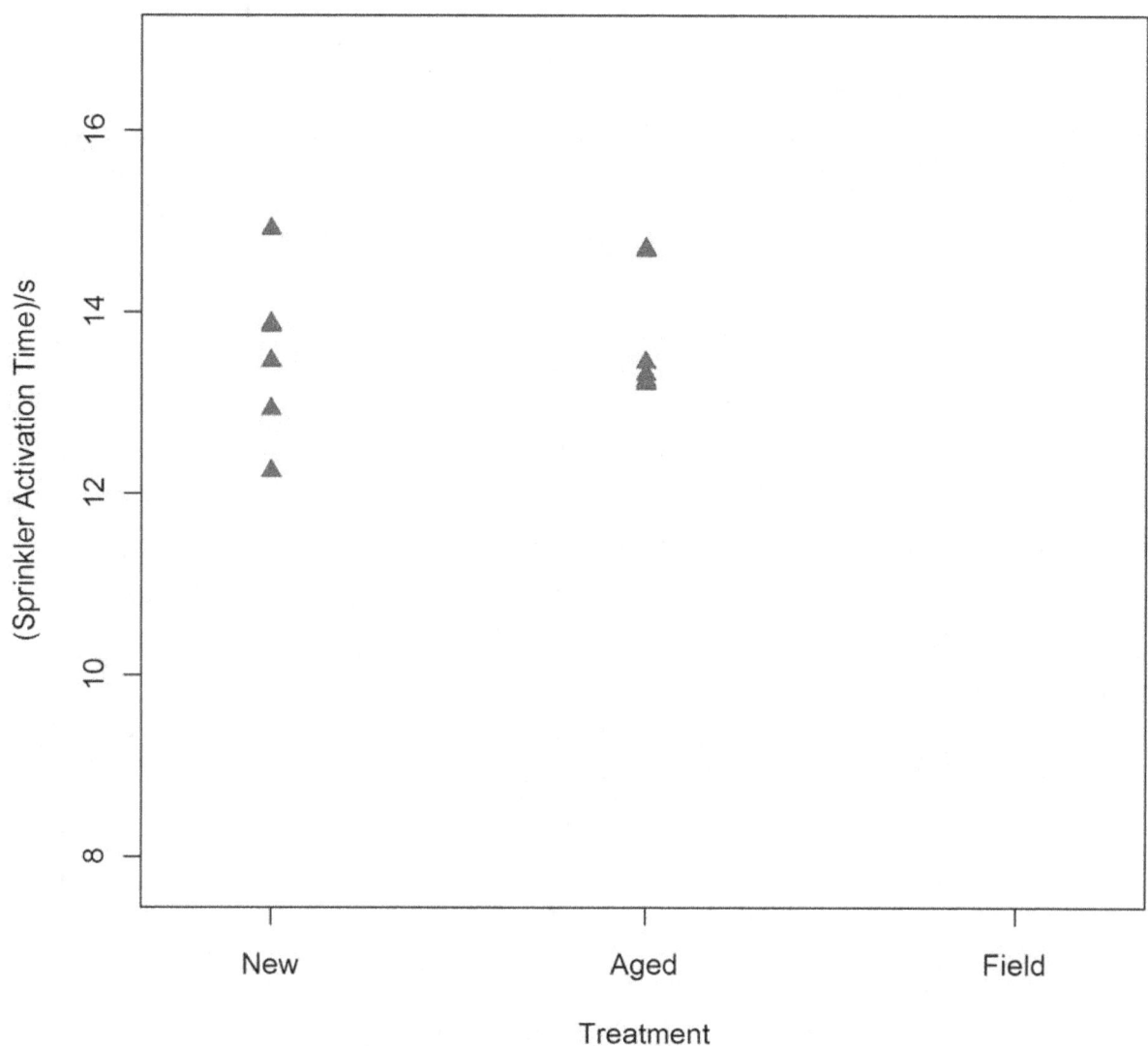

Figure B.3: Sprinkler activation times versus treatment for bulb-type sprinklers made by Manufacturer 1 (Sets 2 and 3). The response times here appear comparable for both new sprinklers and those subjected to the accelerated aging process at Sandia National Laboratories. While the variability in the data looks slightly larger for the new detectors relative to the aged, the difference is not likely to be large enough to be statistically significant. No sprinklers of this type and manufacturer were sampled from the field.

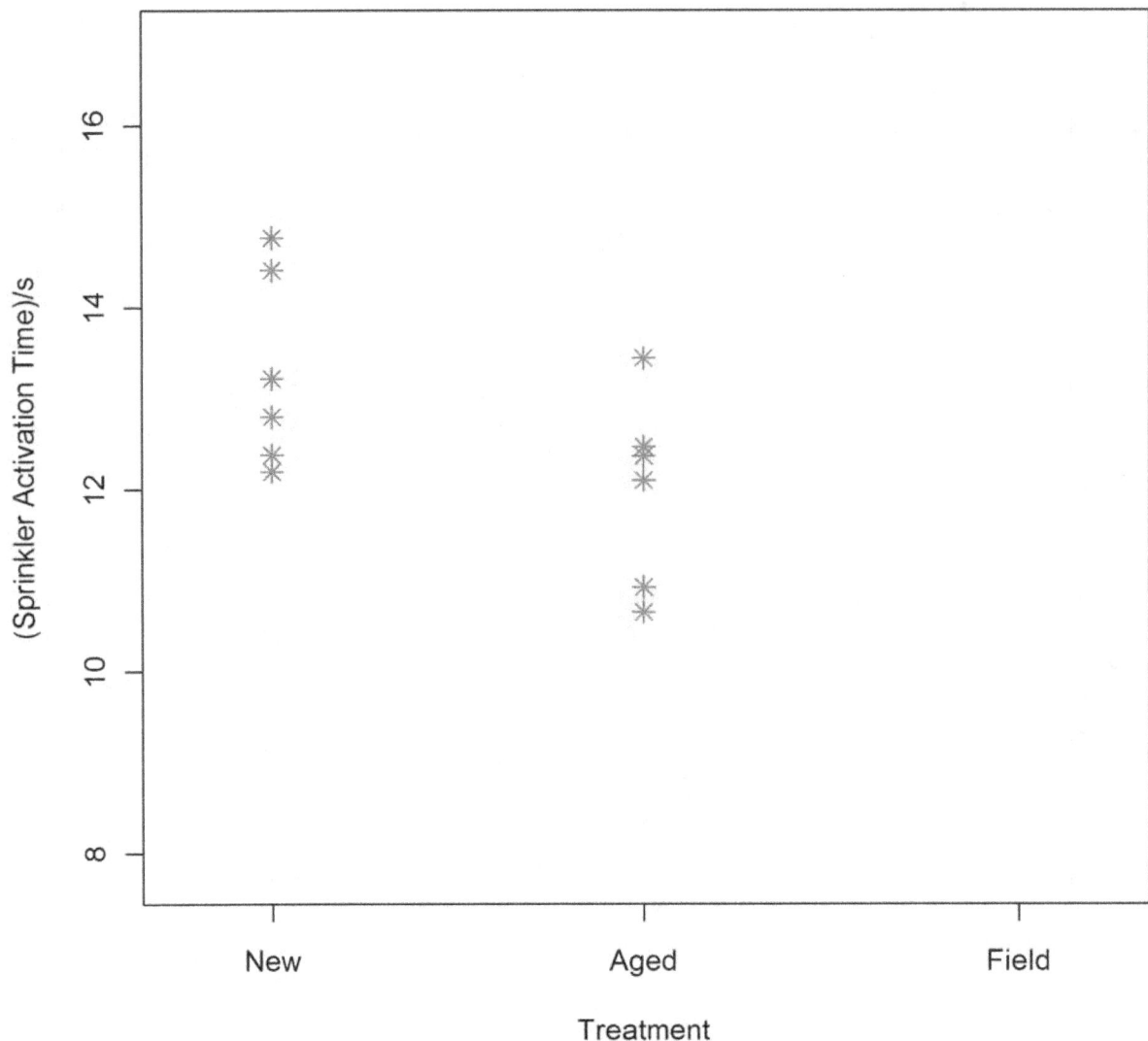

Figure B.4: Sprinkler activation times versus treatment for bulb-type sprinklers made by Manufacturer 2 (Sets 2 and 3). The response times shown here for the sprinklers subjected to the accelerated aging process at Sandia National Laboratories appear to be somewhat lower than those observed for the new sprinklers. The variability in the data for the two treatments looks about the same, and the two low points, when seen in this context, do not appear to be inconsistent with the other aged sprinklers from Manufacturer 2, given the amount of random variation observed between sprinklers within each treatment. No sprinklers of this type and manufacturer were sampled from the field.

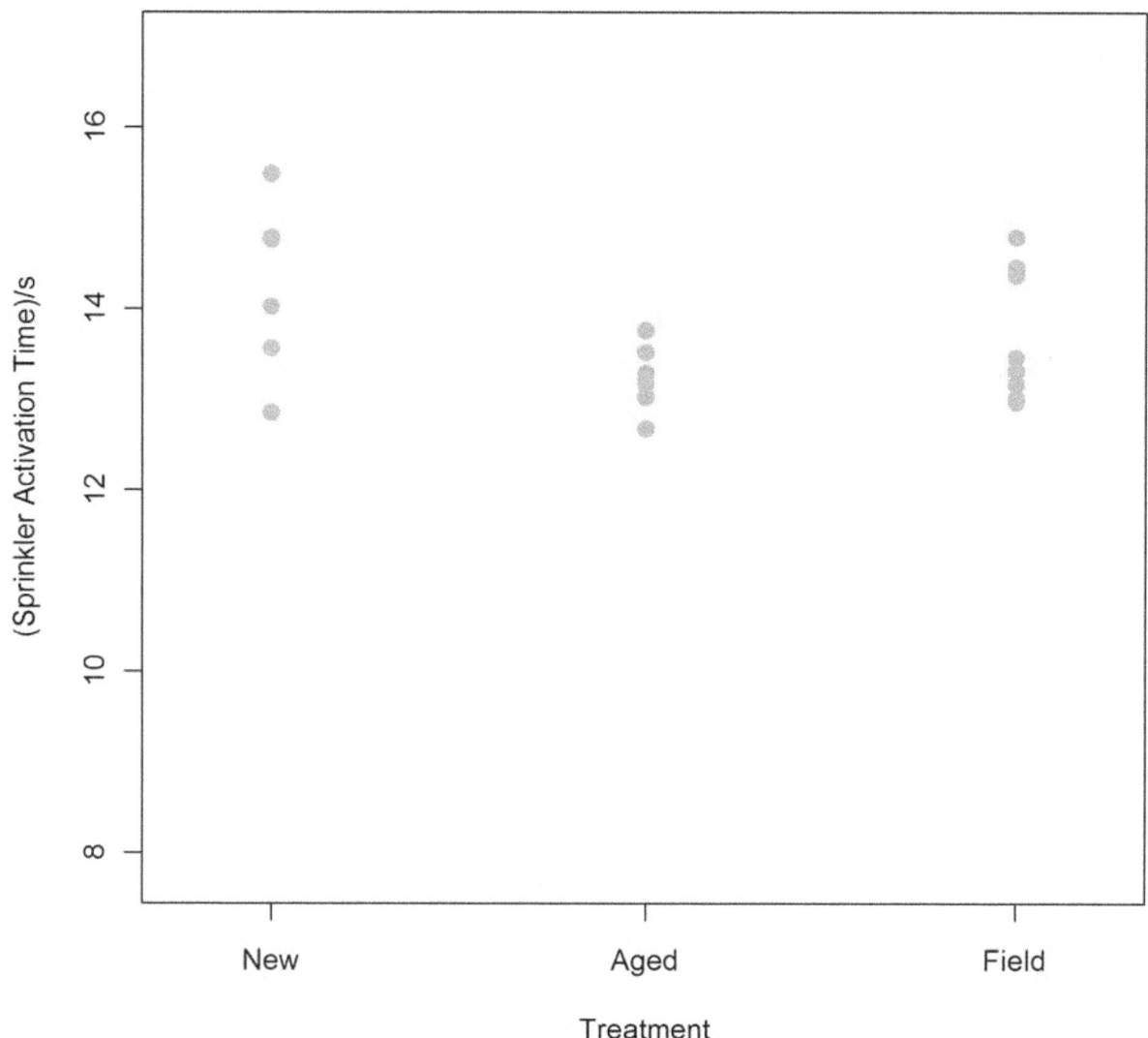

Figure B.5: Sprinkler activation times versus treatment for bulb-type sprinklers made by Manufacturer 3 (Sets 1, 2, and 3). The response times here appear to be fairly similar for all three treatments, although the mean activation time may be slightly lower for the sprinklers subjected to accelerated aging at Sandia National Laboratories. The random variation in the aged and field sprinkler results also appears to be a little less than in the analogous results for new sprinklers. Again, however, these differences are not likely to be large enough to be truly significant from a statistical perspective.

This page intentionally left blank.

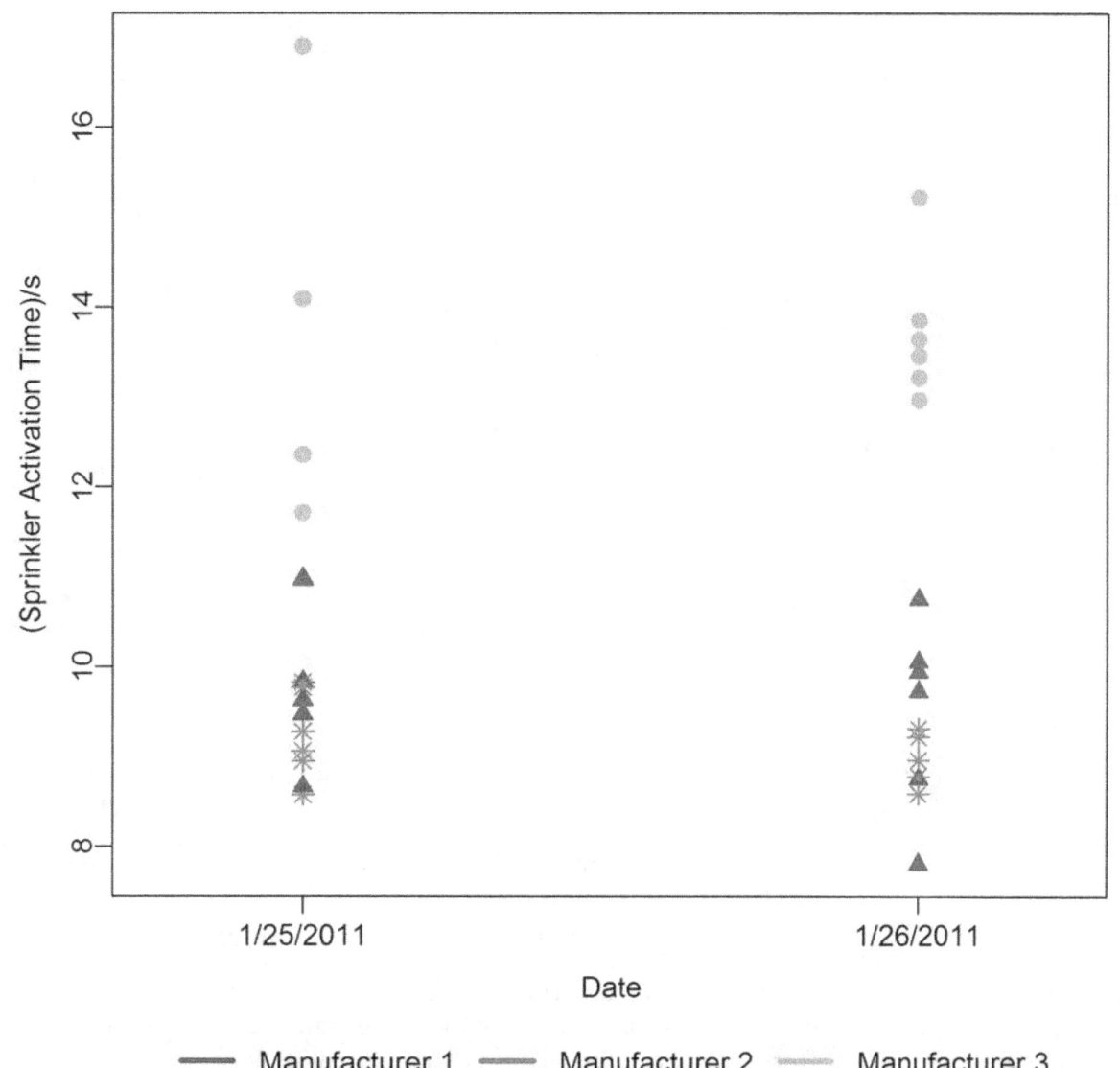

Figure C.1: Sprinkler activation times plotted versus test date for fusible-type sprinklers (Sets 2 and 3). Sprinkler manufacturers are indicated by color coding. This plot shows that the activation times are fairly similar on both test dates, but clearly shows a difference in activation times between Manufacturers 1 and 2 and Manufacturer 3. The random variation in the results for Manufacturer 3 also looks somewhat higher for new sprinklers relative to those subject to the accelerated aging process at Sandia National Laboratories.

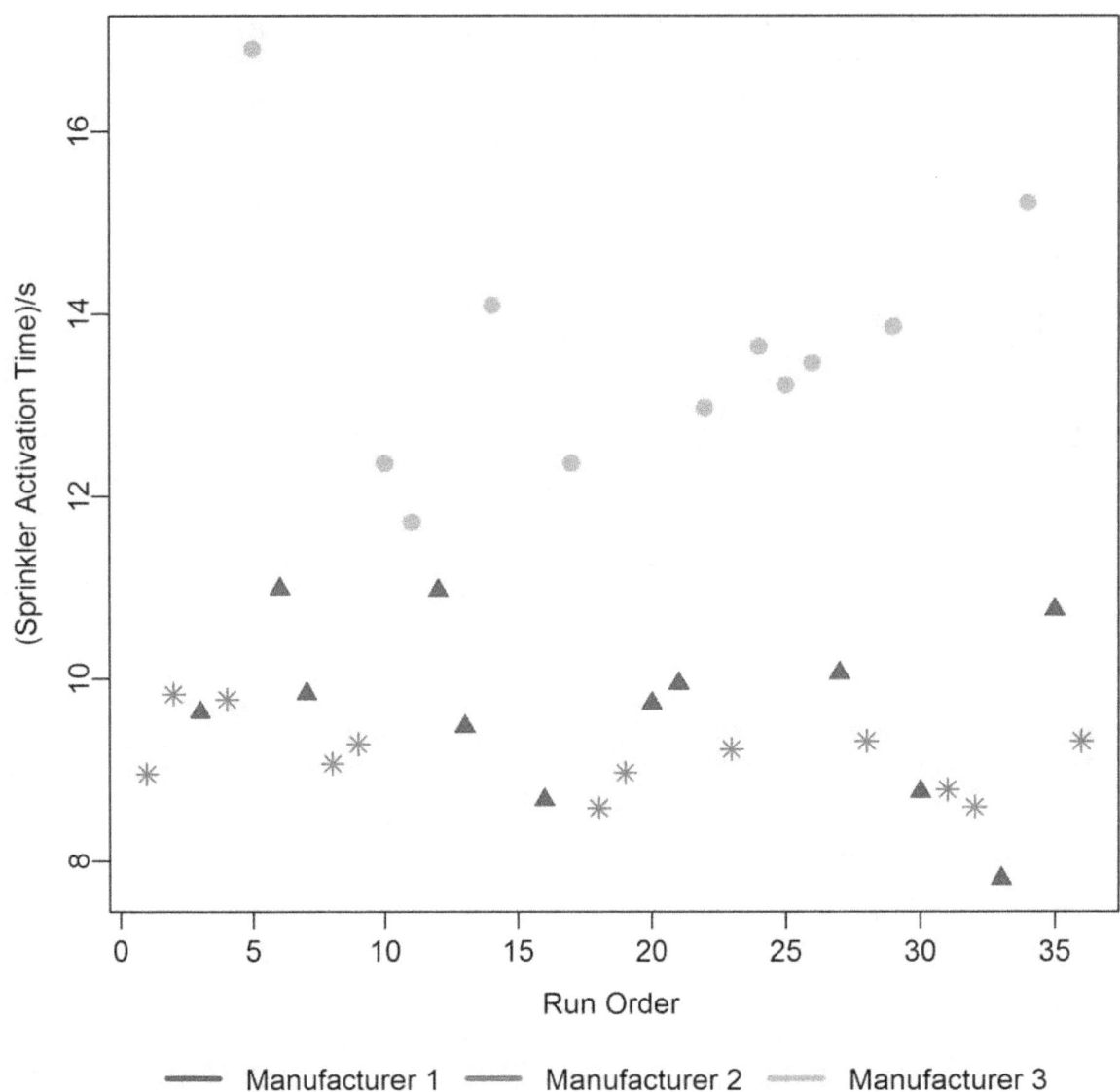

Figure C.2: Sprinkler activation times versus run order for fusible-type sprinklers (Sets 2 and 3). No strong drift over time is apparent in this plot for any of the three manufacturers, although the data for Manufacturer 3 is scattered in such a way that there may be some appearance of drift on first glance.

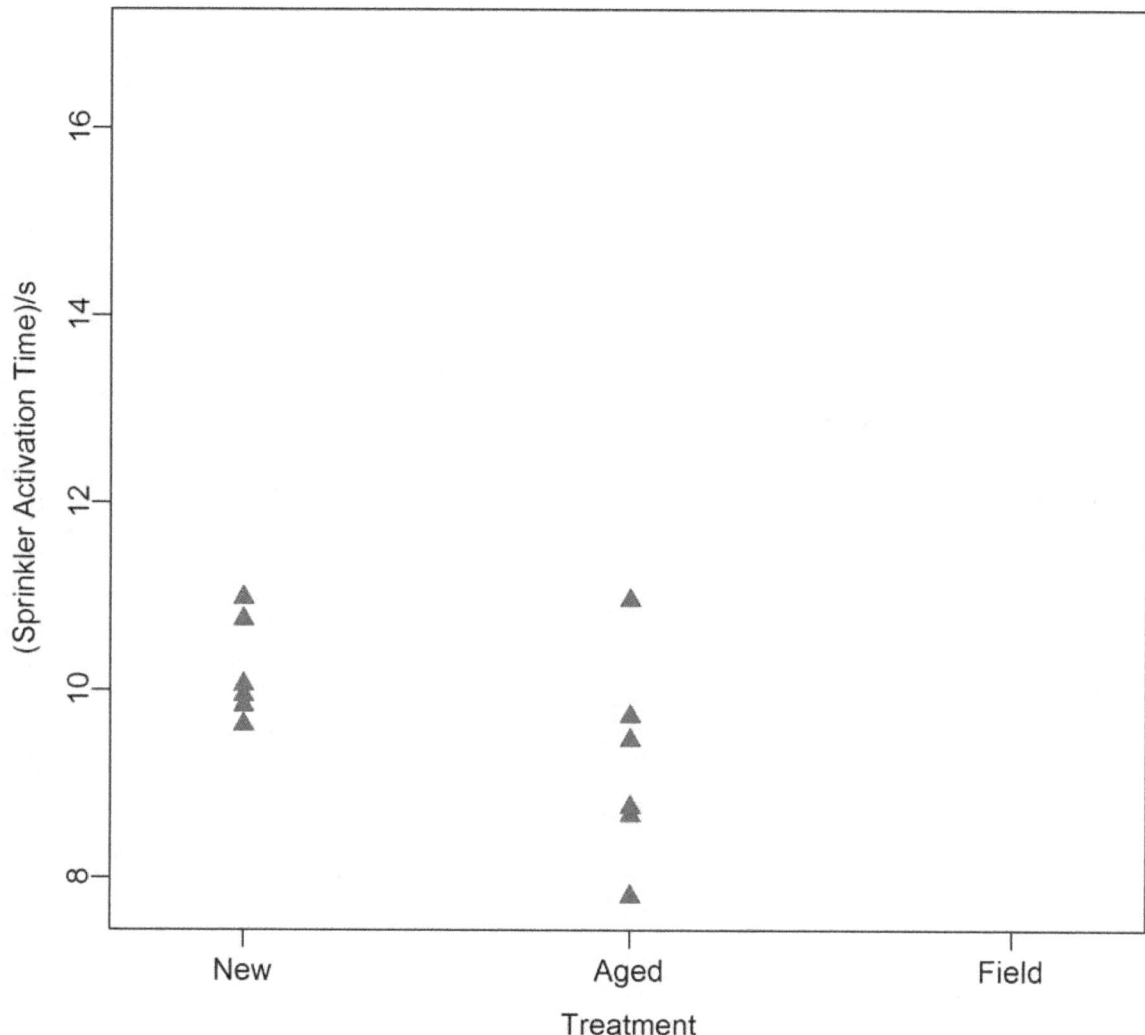

Figure C.3: Sprinkler activation times versus treatment for fusible-type sprinklers made by Manufacturer 1 (Sets 2 and 3). The response times appear comparable for both new sprinklers and those subjected to the accelerated aging process at Sandia National Laboratories. While the variability in the data looks slightly larger for the aged sprinklers relative to the new sprinklers, the difference is not likely to be large enough to be significant. No fusible-type sprinklers were sampled from the field.

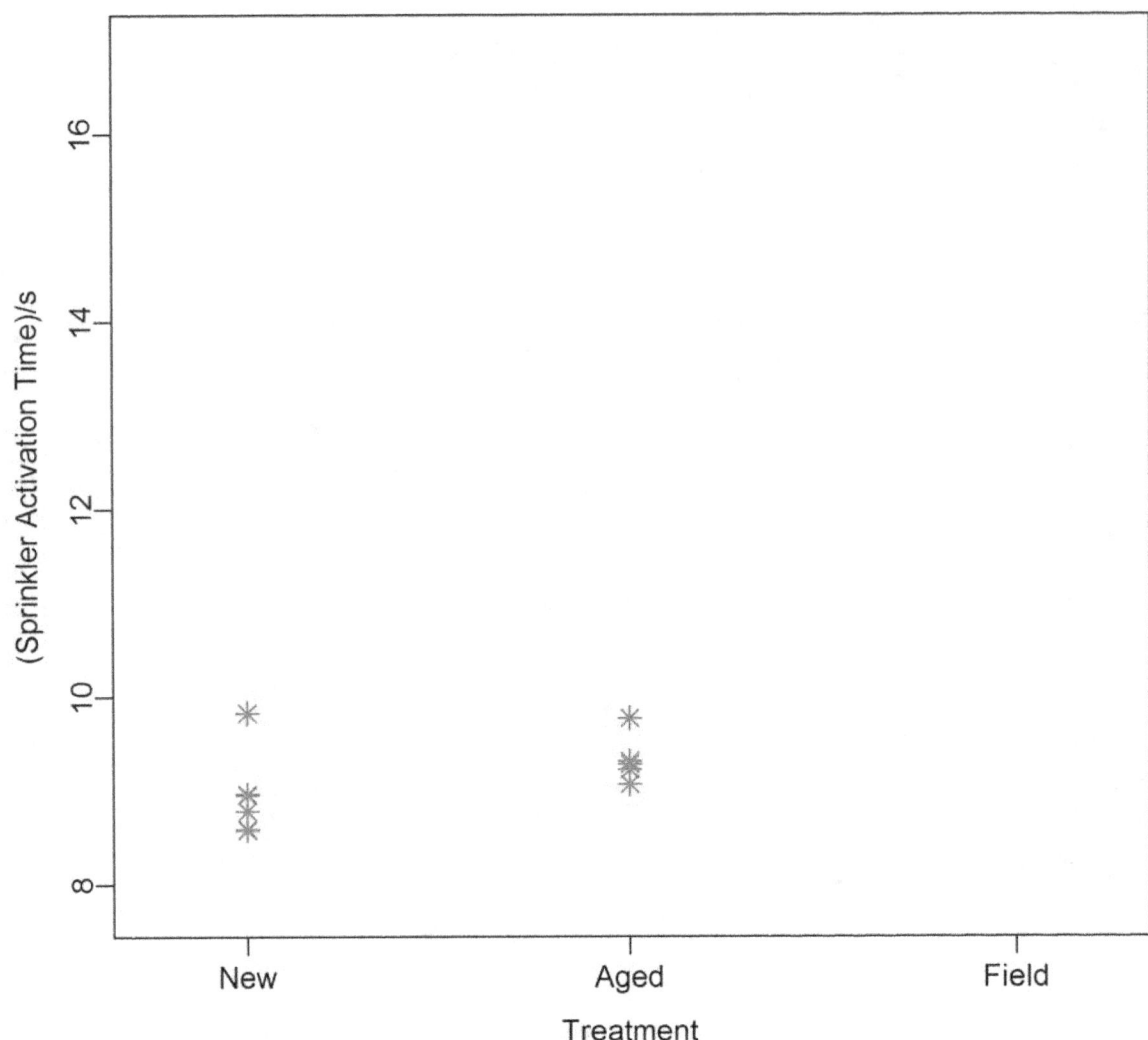

Figure C.4: Sprinkler activation times versus treatment for fusible-type sprinklers made by Manufacturer 2 (Sets 2 and 3). The response times shown here look fairly consistent between treatments. The variability in the data for the two treatments looks about the same as well. No fusible type sprinklers were sampled from the field.

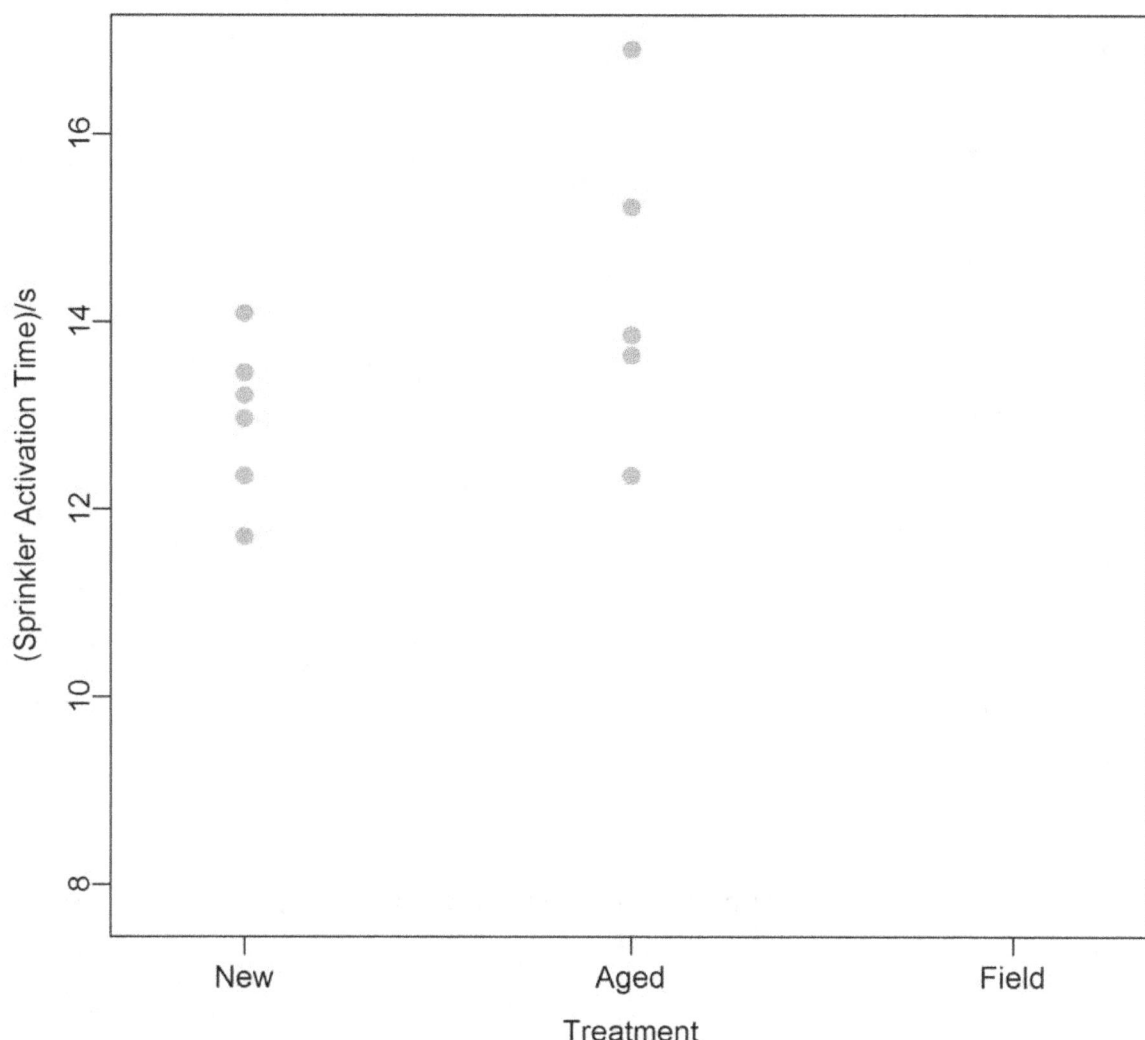

Figure C.5: Sprinkler activation times versus treatment for fusible-type sprinklers made by Manufacturer 3 (Sets 2 and 3). The response times here appear to be slightly higher for the sprinklers subjected to accelerated aging at Sandia National Laboratories than for those tested when new. The random variation in the aged sprinkler results also appears to be a little larger than in the analogous results for new sprinklers. Again, however, these differences are not statistically significant. No fusible type fusible-type sprinklers were sampled from the field.

This page intentionally left blank.

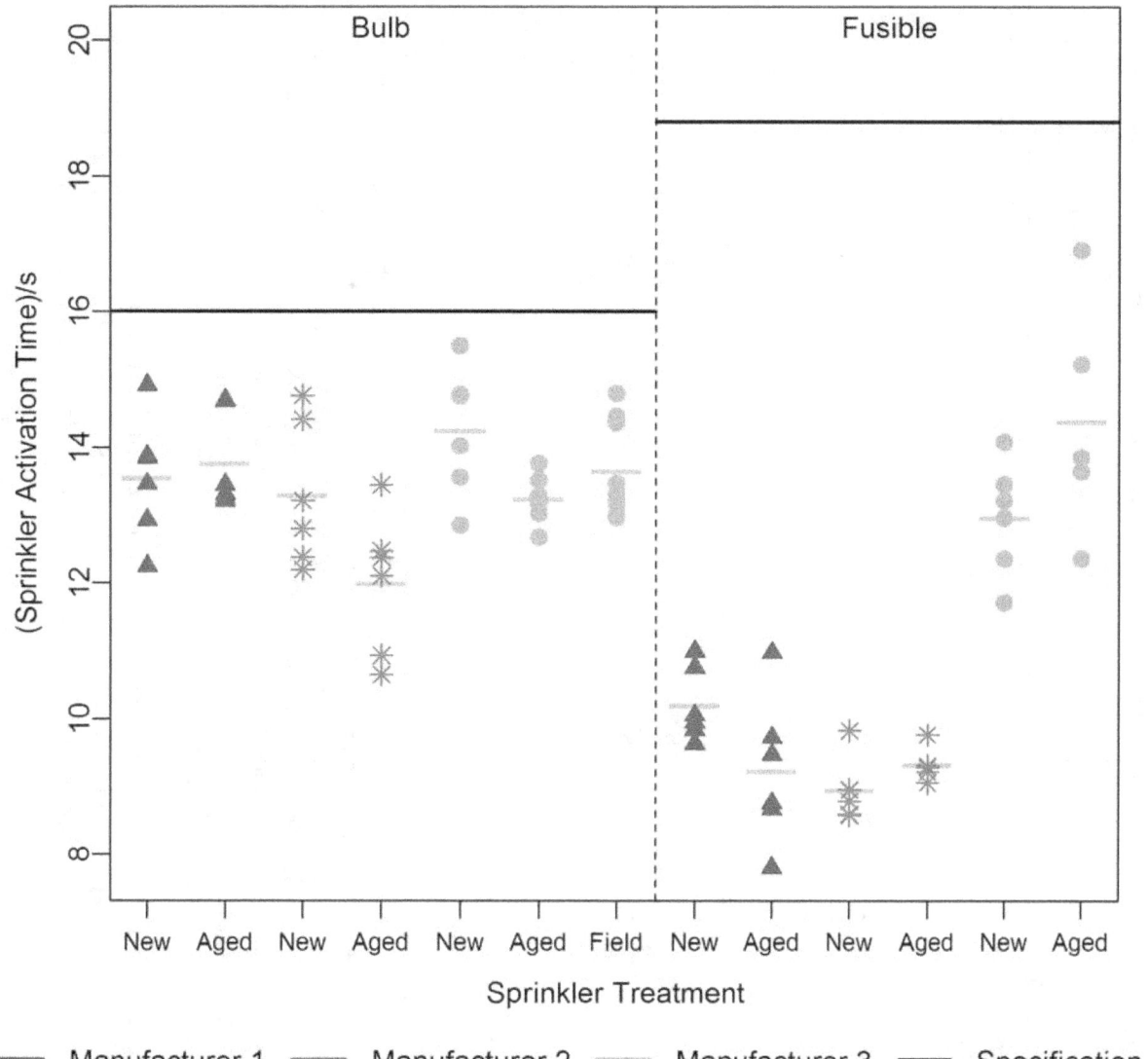

Figure D.1: Sprinkler activation times plotted versus sprinkler treatment. The left-hand section of the plot shows the results for bulb-type sprinklers, while the right-hand section shows the results for fusible-type sprinklers. The black reference lines at 16.0 s and 18.8 s are the performance specifications for the activation times for each type of sprinkler. The gray line in the middle of each set of observed activation times is the mean activation time for that set of data. All bulb-type sprinklers activated prior to their specification of 16.0 s. One of the fusible-type sprinklers made by Manufacturer 3 and aged at Sandia National Laboratories did not activate after 600 s of exposure in the plunge test. All other fusible sprinklers activated before, or well before, the specified maximum activation time of 18.8 s.

Sprinkler Operating Principle	Mfr	Treatment	Mean Activation Time[b] (s)	Std. Dev. Activation Time (s)	Number of Sprinklers Tested	Pooled Std. Dev. (s)	Degrees of Freedom for Pooled Std. Dev.	One-Sided Coverage Factor 95 % Confidence 99 % Content	95 %/99 % Upper Tolerance Limit (s)	Specified Activation Time (s)
Bulb	1	New	13.55	0.91	6	0.82	10	3.96	16.79	16.0
Bulb	1	Aged	13.77	0.72	6	0.82	10	3.96	17.01	16.0
Bulb	2	New	13.30	1.07	6	1.05	10	3.96	17.47	16.0
Bulb	2	Aged	12.00	1.04	6	1.05	10	3.96	16.17	16.0
Bulb	3	New	14.24	0.95[a]	6	0.71[a]	18	3.53	16.76	16.0
Bulb	3	Aged	13.24	0.38[a]	6	0.71[a]	18	3.53	15.75	16.0
Bulb	3	Field	13.65	0.69[a]	9	0.71[a]	18	3.44	16.09	16.0
Fusible	1	New	10.20	0.54	6	0.86	10	3.96	13.60	18.8
Fusible	1	Aged	9.23	1.09	6	0.86	10	3.96	12.63	18.8
Fusible	2	New	8.95	0.46	6	0.37	10	3.96	10.40	18.8
Fusible	2	Aged	9.33	0.24	6	0.37	10	3.96	10.78	18.8
Fusible	3	New	12.96	0.84	6	1.31	9	4.07	18.31	18.8
Fusible	3	Aged	14.39	1.73	5	1.31	9	4.12	19.80	18.8

[a] As described in Appendix A, these values were not computed with a variance component for day-to-day variability, although a small, statistically significant effect was observed. To ensure that this simplification has not skewed the results, tolerance limits were also computed based only on the data from 1/31/2011, the day with the longer activation times on average. Using the only the values from the day with the longer average times is a relatively conservative approach to assessing these tolerance limits in the absence of a precise assessment of the magnitude of the typical day-to-day variations in these results. The results from the data collected on 1/31/2011 are: New: mean = 15.01 s, tolerance limit = 17.01 s, 3 units tested; Aged: mean = 13.36 s, tolerance limit = 15.36 s, 3 units tested; Field: mean = 14.08 s, tolerance limit = 15.98 s, 5 units tested. This data was used to perform statistical analysis, in order to prevent rounding errors. The data presented in the table is rounded to 1/100 s due to the plunge test activation time measurement uncertainties calculated in Section IV.A.B.

[b] Activation times were reported to 1/1000 s by the plunge test apparatus.

Table D.1: 95 % confidence 99 % content upper statistical tolerance limits [18] for sprinkler activation times (further details in Appendix A). These limits provide a bound that will exceed the activation times of 99 % of all future sprinklers from each population with 95 % confidence. The fact that several of these limits exceed their specifications simply indicates this data does not prove the sprinklers all meet specifications. With more data, however, the bounds would be tighter on average because the coverage factor for these bounds [18] is a decreasing function of the number of measurements. Thus, with more data, a smaller tolerance limit could indicate that the sprinklers do meet specifications.

44

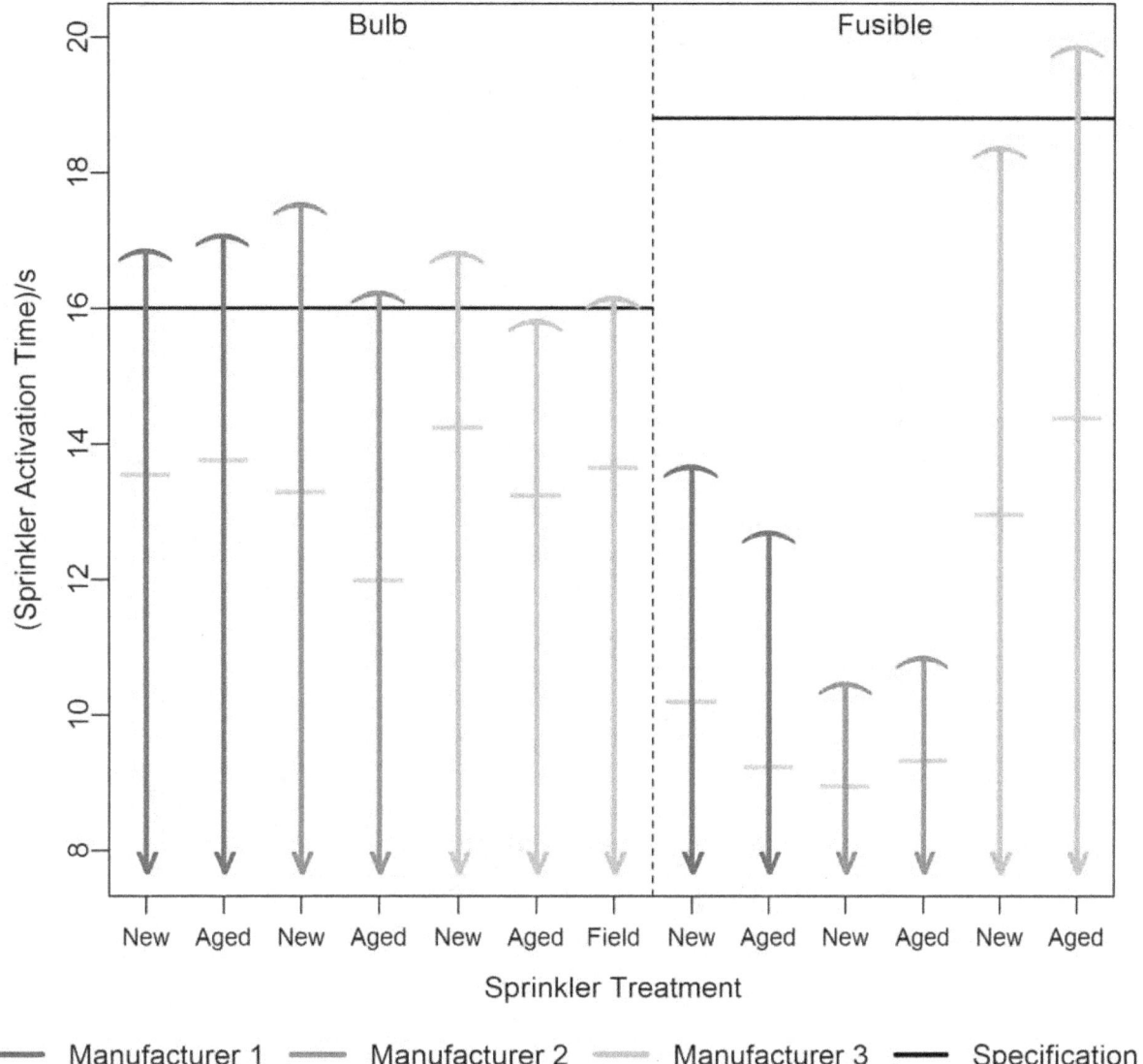

Figure D.2: 95 % confidence 99 % content upper statistical tolerance limits for sprinkler activation times from Table D.1. The curved upper bound shown for each population of sprinkler activation times will exceed the activation times of 99 % of all sprinklers from the associated population with 95 % confidence. The gray line associated with each tolerance limit is the sample mean of the observed activation times for the associated set of data. This plot shows that for all sprinkler populations studied, the upper bound is near the specified maximum activation time.

This page intentionally left blank.

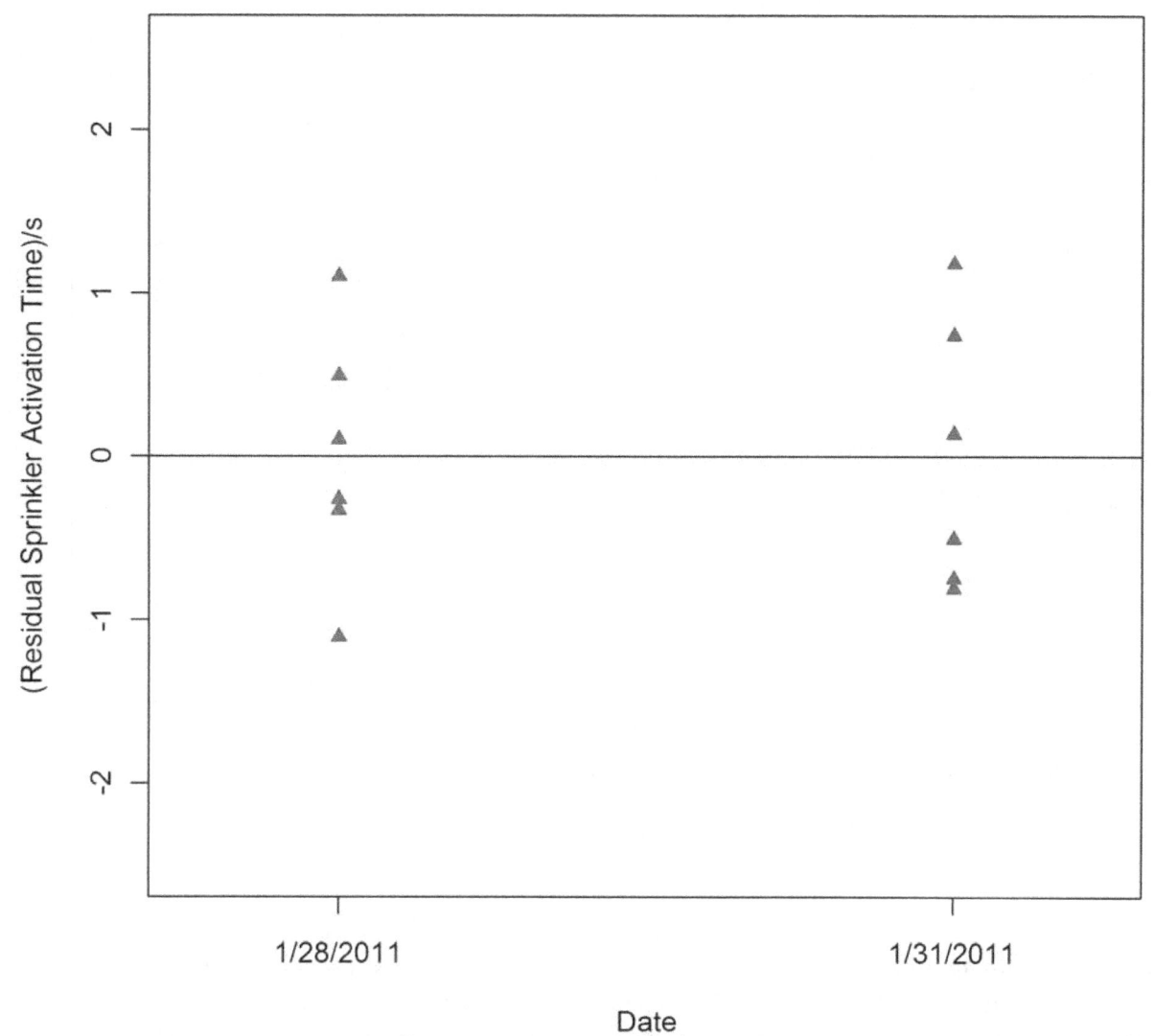

Figure E.1: Residuals from the fit of the model described in Appendix A to sprinkler activation times for bulb-type sprinklers made by Manufacturer 1 (Sets 2 and 3) plotted versus sprinkler test date. This plot indicates that any significant test date effects have been successfully accounted for by the model since the residuals for each test date are centered on a mean value of 0 and display essentially the same level of random variation.

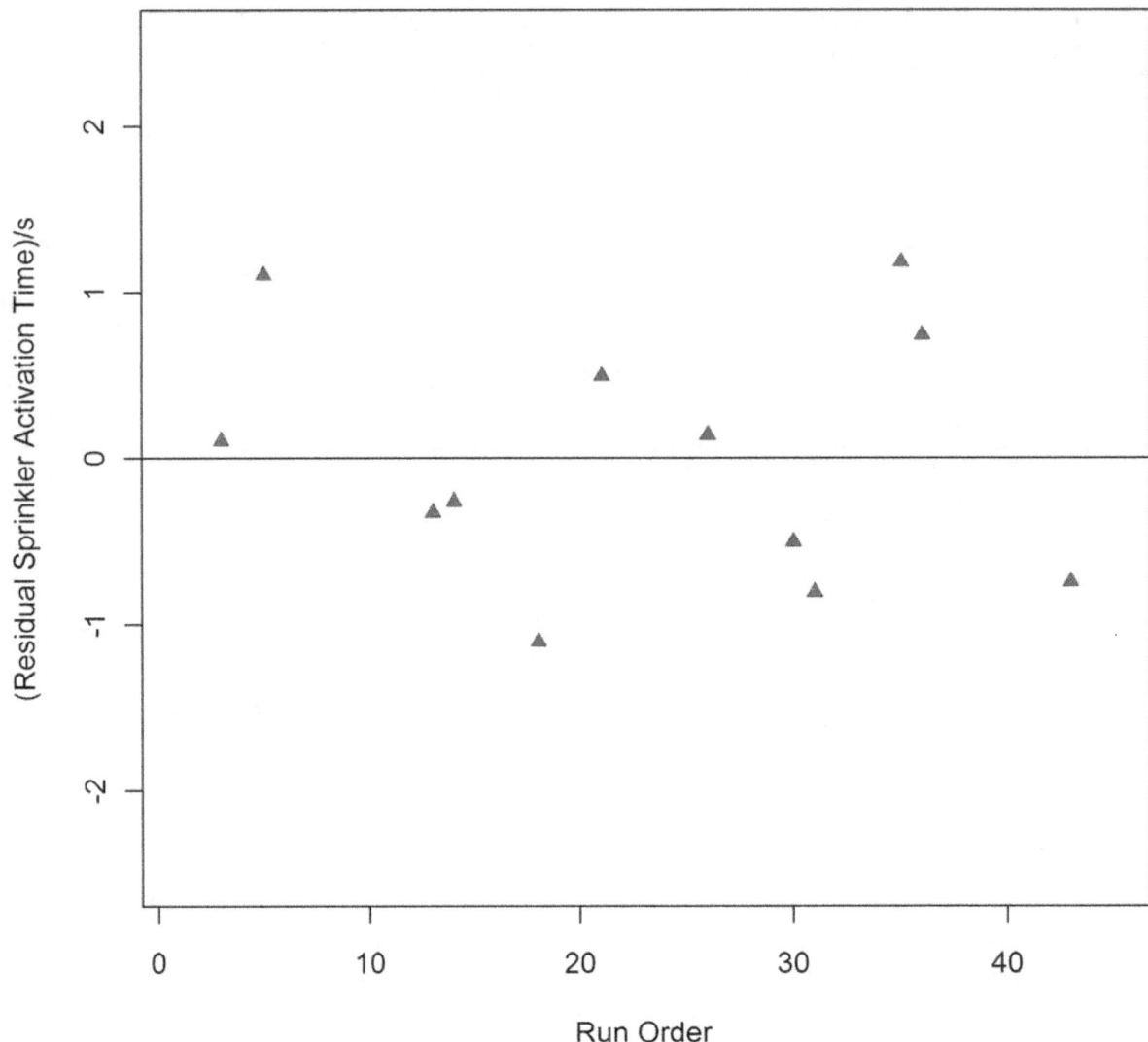

Figure E.2: Residuals from the fit of the model described in Appendix A to sprinkler activation times for bulb-type sprinklers made by Manufacturer 1 (Sets 2 and 3) plotted versus run order. This plot does not identify any drift in the measurements over time because the residuals appear to be scattered randomly around a mean value of 0 and display essentially the same level of random variation across all runs over time.

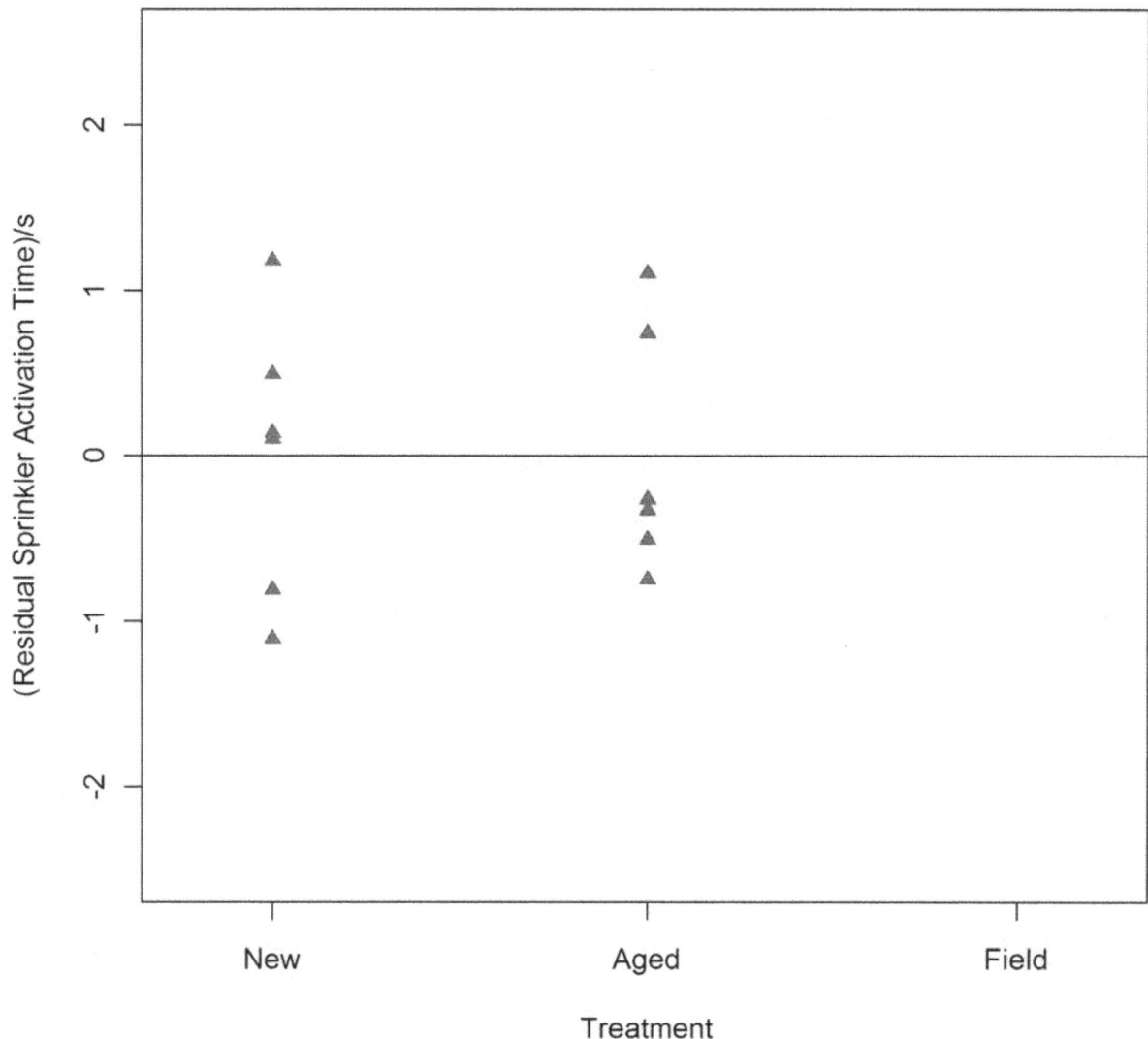

Figure E.3: Residuals from the fit of the model described in Appendix A to sprinkler activation times for bulb-type sprinklers made by Manufacturer 1 (Sets 2 and 3) plotted versus treatment. This plot indicates that any significant treatment effects have been successfully accounted for by the model since the residuals for each test date are centered on a mean value of 0 and display essentially the same level of random variation. No sprinklers of this type and manufacturer were sampled from the field.

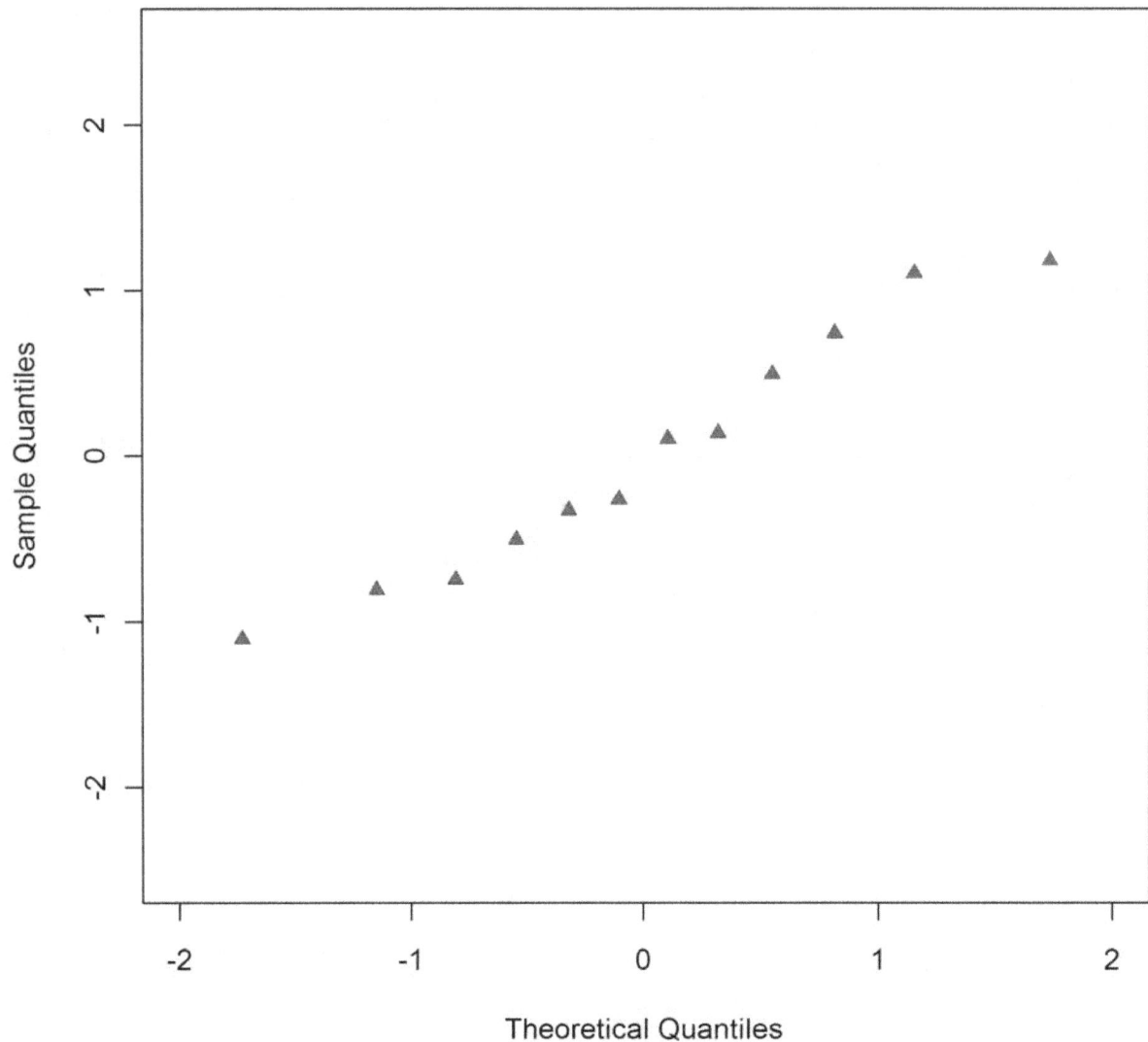

Figure E.4: Normal probability plot of the residuals from the fit of the model described in Appendix A to sprinkler activation times for bulb-type sprinklers made by Manufacturer 1 (Sets 2 and 3). The fact that the residuals essentially fall along a straight line when plotted versus theoretical quantiles from the standard normal distribution suggests that the random errors follow a distribution that reasonably approximates the assumed normal distribution.

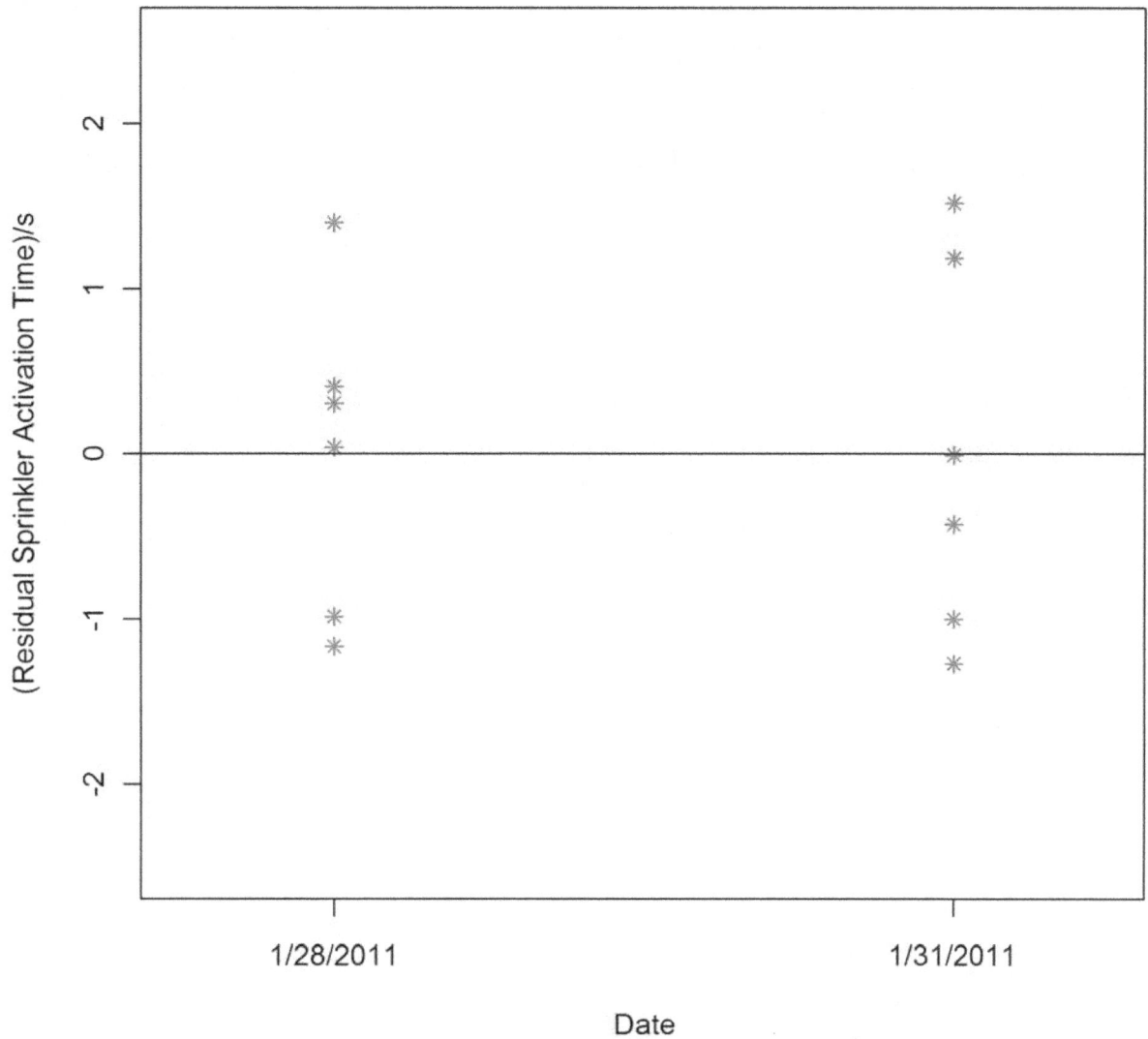

Figure E.5: Residuals from the fit of the model described in Appendix A to sprinkler activation times for bulb-type sprinklers made by Manufacturer 2 (Sets 2 and 3) plotted versus sprinkler test date. This plot indicates that any significant test date effects have been successfully accounted for by the model since the residuals for each test date are centered on a mean value of 0 and display essentially the same level of random variation.

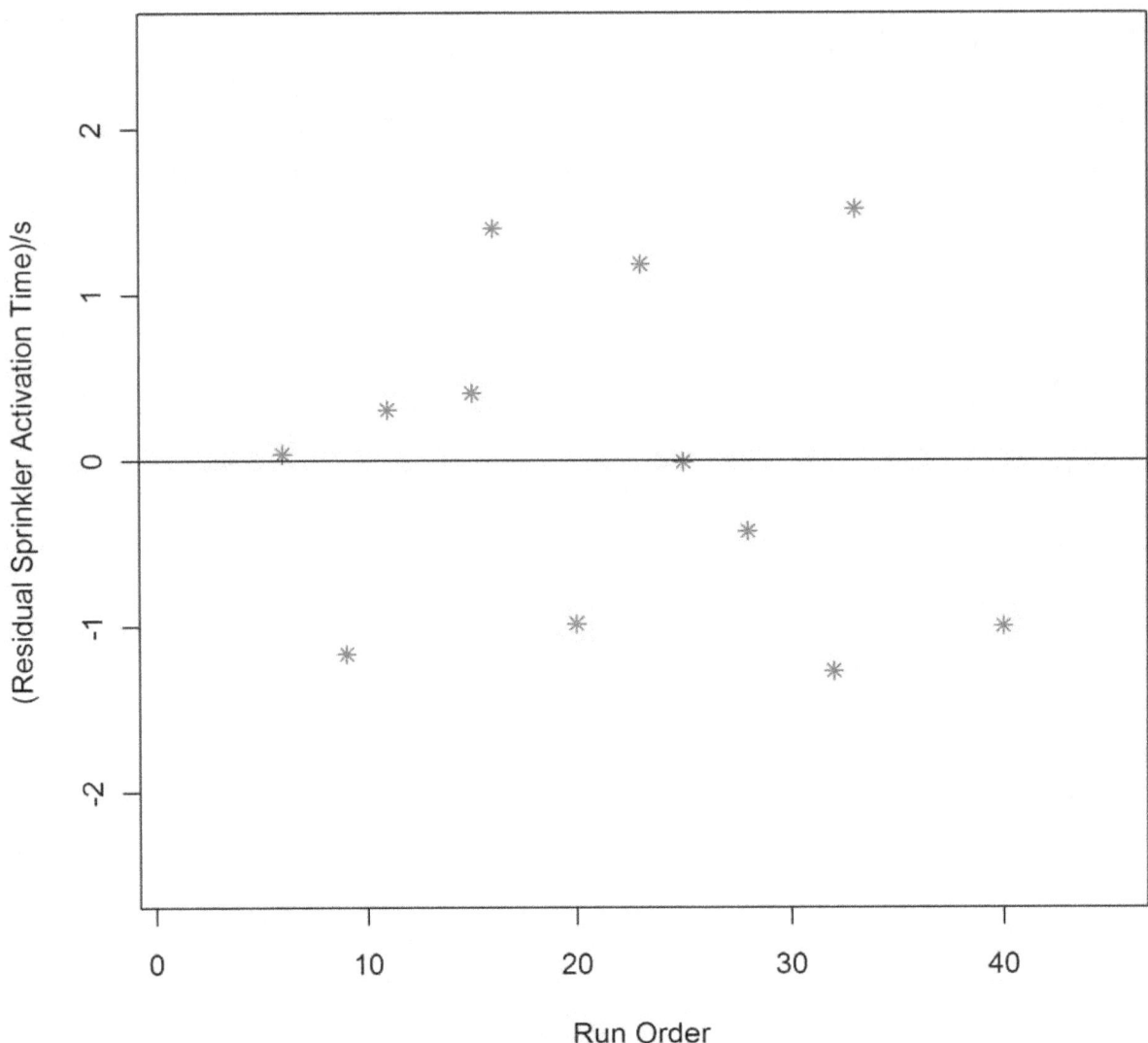

Figure E.6: Residuals from the fit of the model described in Appendix A to sprinkler activation times for bulb-type sprinklers made by Manufacturer 2 (Sets 2 and 3) plotted versus run order. This plot does not identify any drift in the measurements over time since the residuals appear to be randomly scattered around a mean value of 0 and display essentially the same level of random variation across all runs over time.

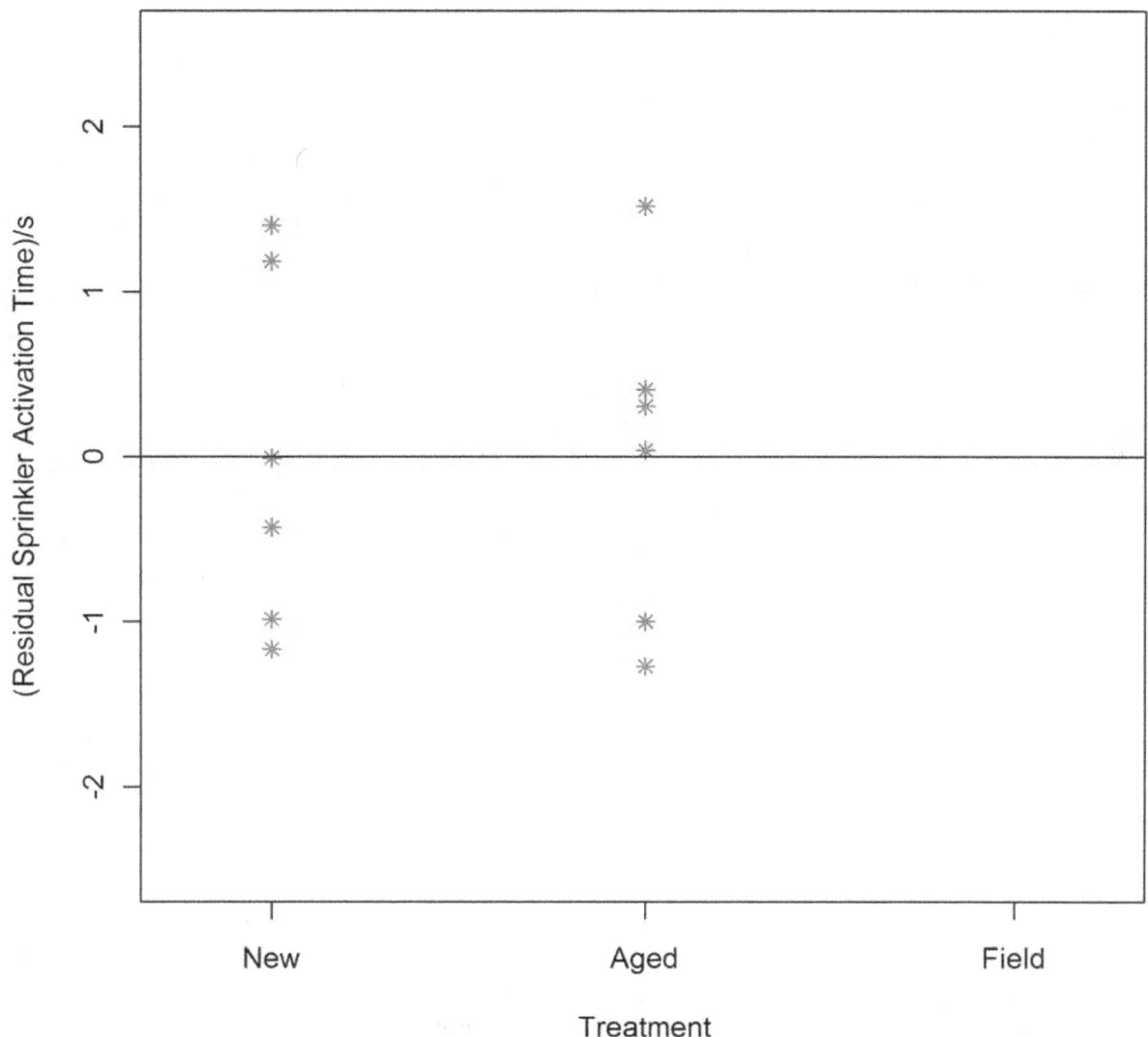

Figure E.7: Residuals from the fit of the model described in Appendix A to sprinkler activation times for bulb-type sprinklers made by Manufacturer 2 (Sets 2 and 3) plotted versus treatment. This plot indicates that any significant treatment effects have been successfully accounted for by the model since the residuals for each test date are centered on a mean value of 0 and display essentially the same level of random variation. No sprinklers of this type and manufacturer were sampled from the field.

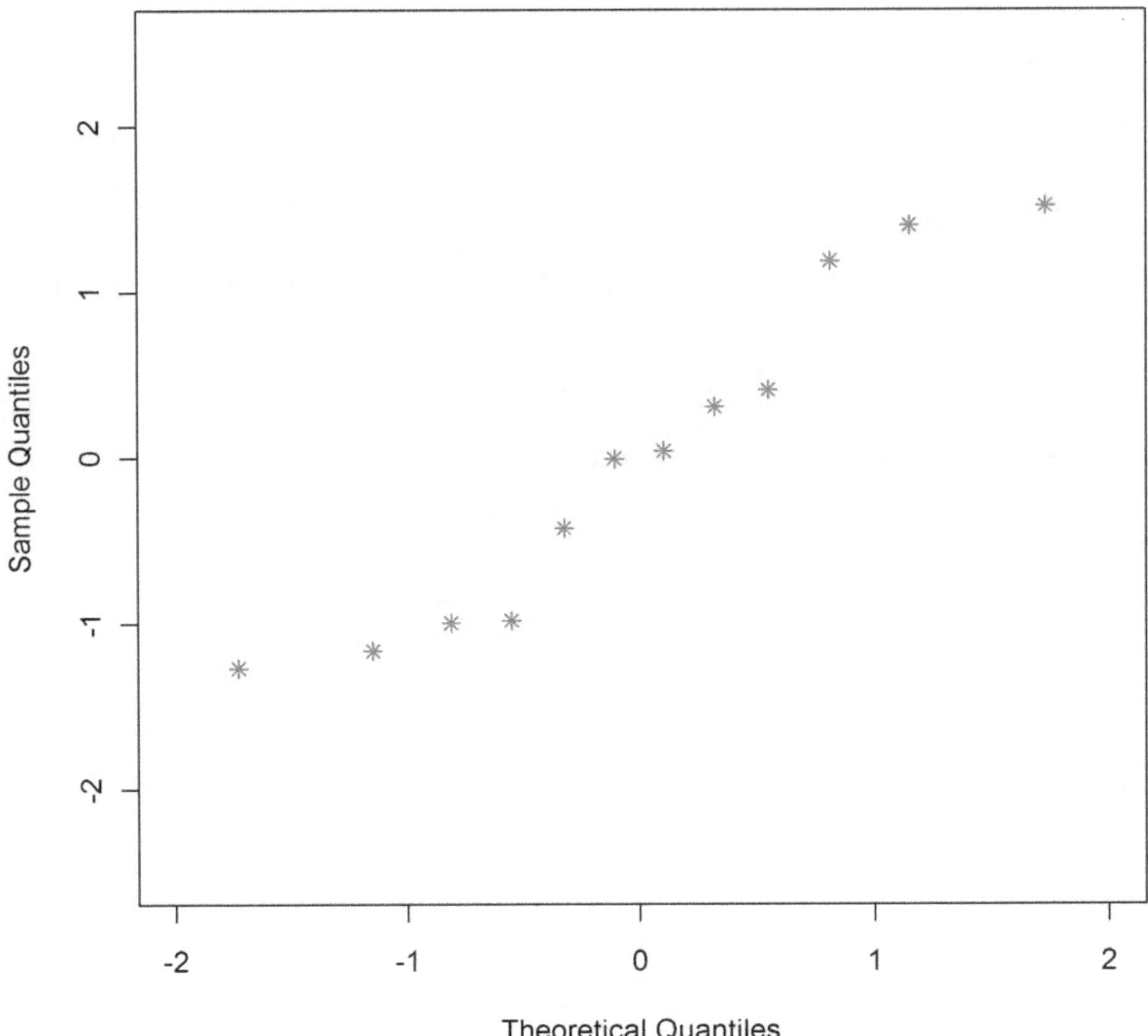

Figure E.8: Normal probability plot of the residuals from the fit of the model described in Appendix A to sprinkler activation times for bulb-type sprinklers made by Manufacturer 2 (Sets 2 and 3). The fact that the residuals essentially fall along a straight line when plotted versus theoretical quantiles from the standard normal distribution suggests that the random errors follow a distribution that reasonably approximates the assumed normal distribution.

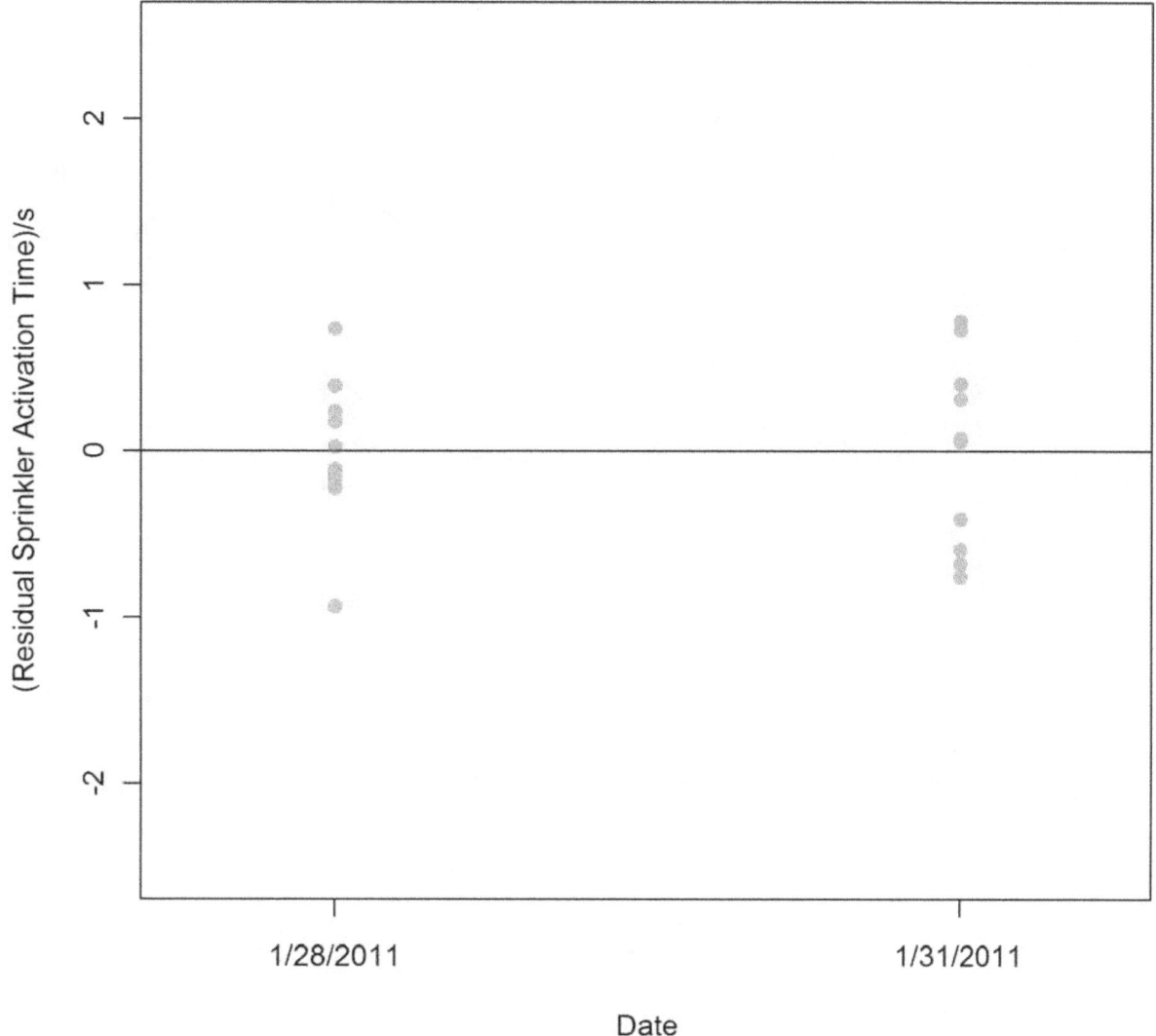

Figure E.9: Residuals from the fit of the model described in Appendix A to sprinkler activation times for bulb-type sprinklers made by Manufacturer 3 (Sets 1, 2, and 3) plotted versus sprinkler test date. This plot indicates that any significant test date effects have been successfully accounted for by the model since the residuals for each test date are centered on a mean value of 0 and display essentially the same level of random variation.

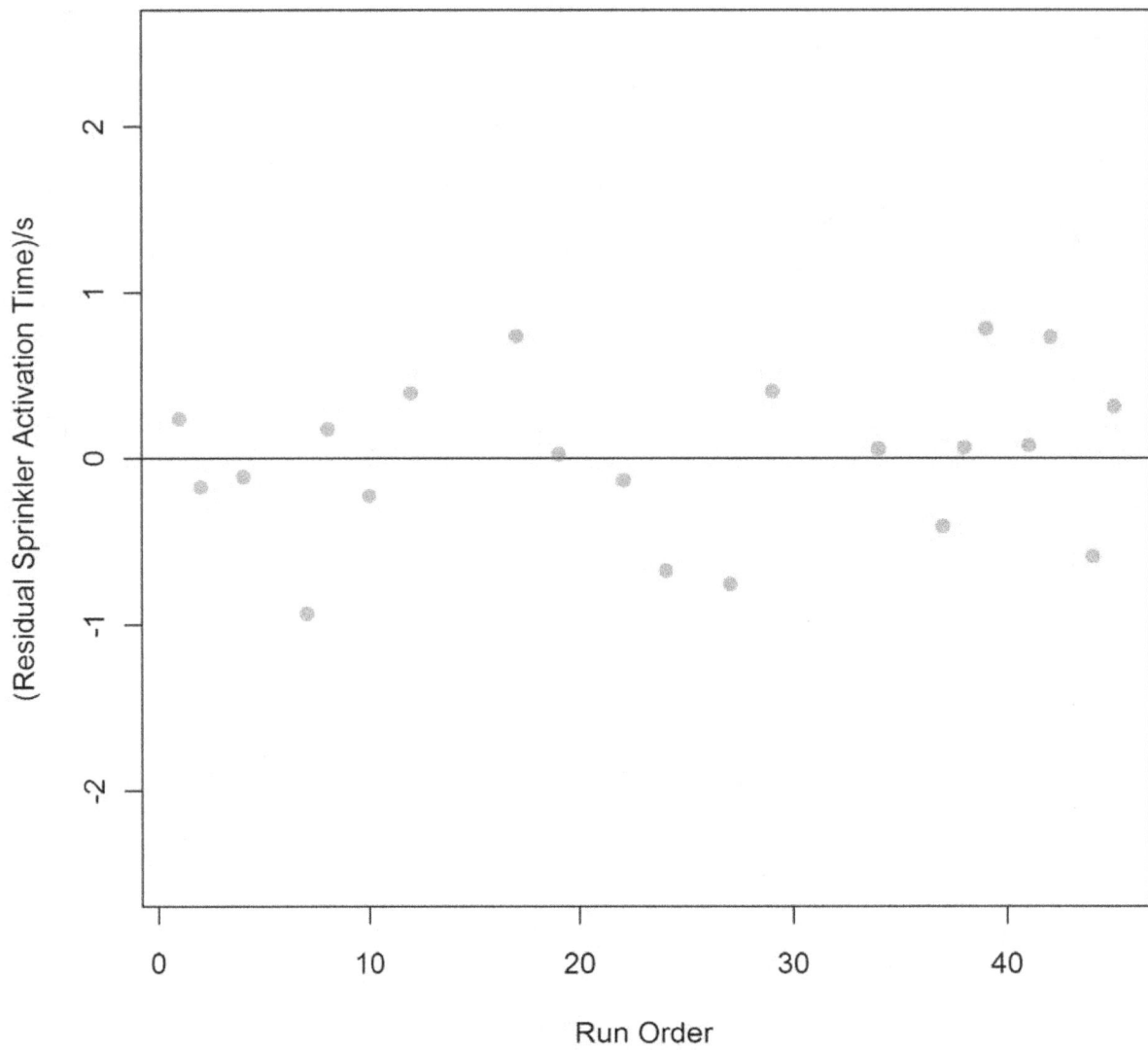

Figure E.10: Residuals from the fit of the model described in Appendix A to sprinkler activation times for bulb-type sprinklers made by Manufacturer 3 (Sets 1, 2, and 3) plotted versus run order. This plot does not identify any drift in the measurements over time since the residuals appear to be randomly scattered around a mean value of 0 and display essentially the same level of random variation across all runs over time.

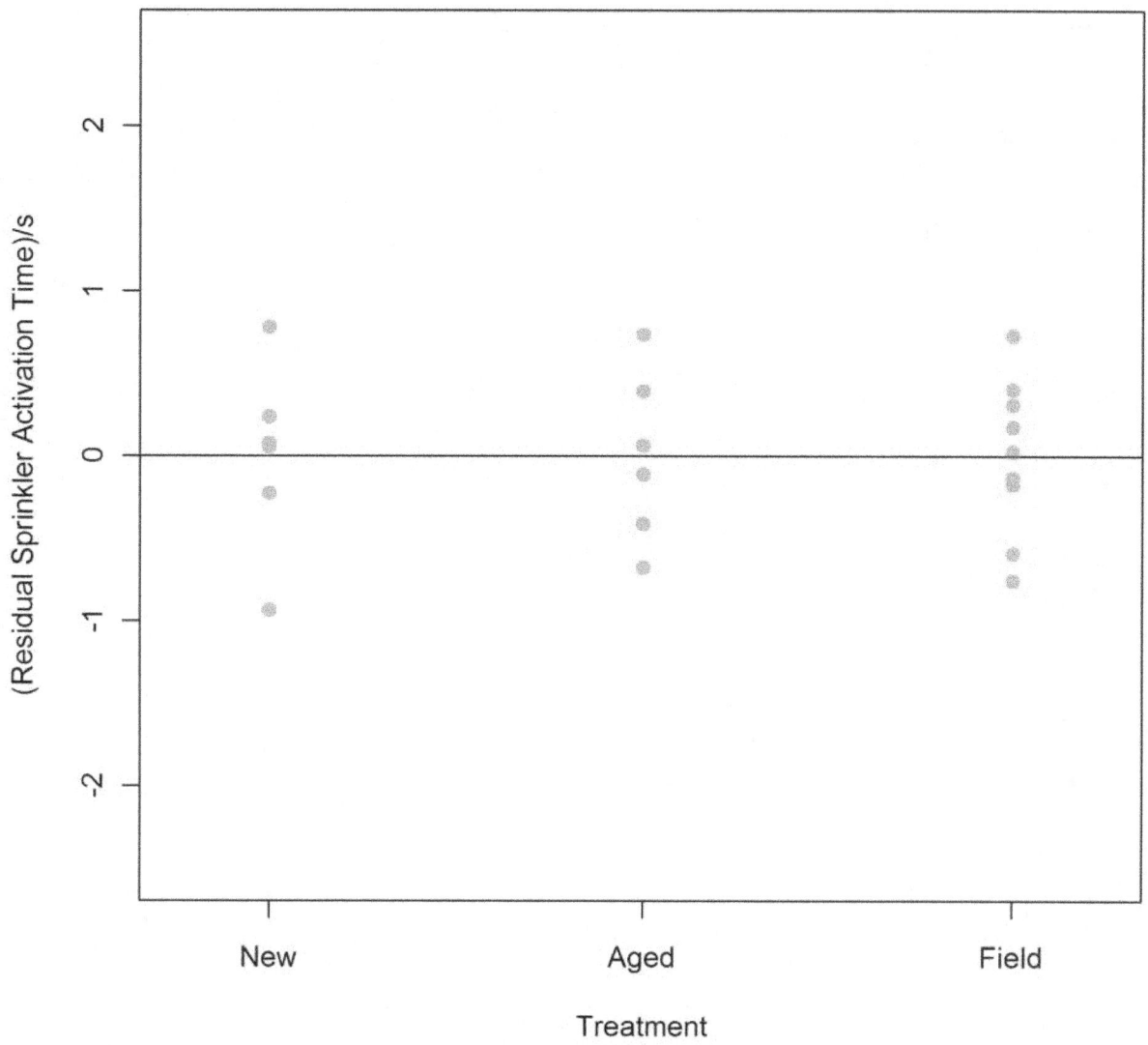

Figure E.11: Residuals from the fit of the model described in Appendix A to sprinkler activation times for bulb-type sprinklers made by Manufacturer 3 (Sets 1, 2, and 3) plotted versus treatment. This plot indicates that any significant treatment effects have been successfully accounted for by the model since the residuals for each test date are centered on a mean value of 0 and display essentially the same level of random variation.

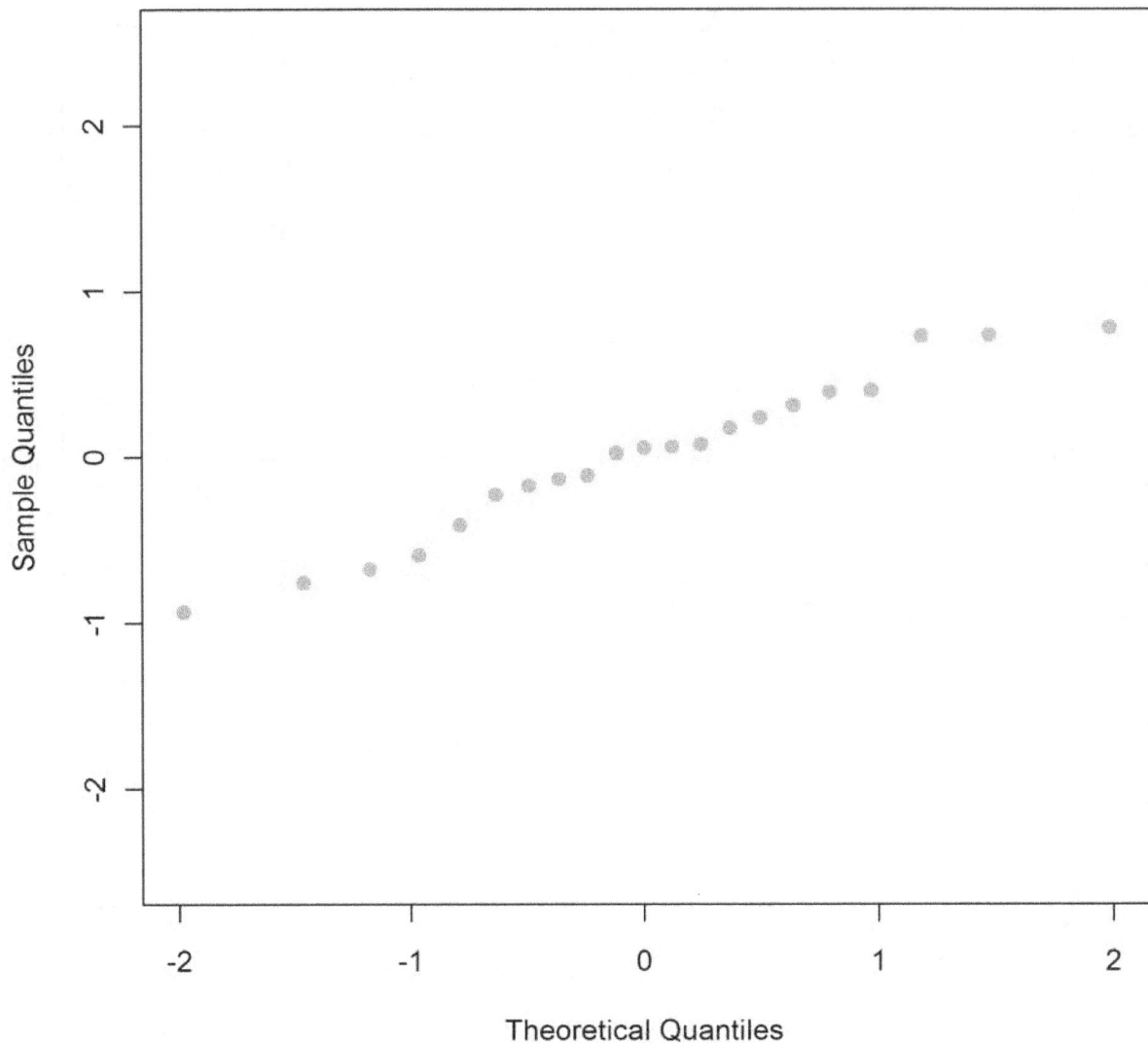

Figure E.12: Normal probability plot of the residuals from the fit of the model described in Appendix A to sprinkler activation times for bulb-type sprinklers made by Manufacturer 3 (Sets 1, 2, and 3). The fact that the residuals essentially fall along a straight line when plotted versus theoretical quantiles from the standard normal distribution suggests that the random errors follow a distribution that reasonably approximates the assumed normal distribution.

ANOVA Output for Manufacturer 1

Operating Principle: Bulb

	Degrees of Freedom	Sum of Squares	Mean Square	F Value	Pr(>F)
Test Date	1	0.431	0.431	0.616	0.453
Treatment	1	0.146	0.146	0.208	0.659
Residuals	9	6.293	0.699		

Table E.1: Analysis of variance table for bulb-type sprinklers made by Manufacturer 1 (Sets 2 and 3) used to test the significance of potential test date and treatment effects on sprinkler activation times. Details of the model fit to the data are given in Appendix A. Because the p-values, shown in the column labeled "**Pr(>F)**" are both greater than the chosen significance level of $\alpha = 0.05$, the conclusion is that neither test date nor treatment have any statistically significant effect on the sprinkler activation times. Note, the p-values reported here apply only to the results for Manufacturer 1 for bulb-type sprinklers and have not been adjusted for the additional tests carried out to assess the performance of sprinklers based on other operating principles or made by other manufacturers.

ANOVA Output for Manufacturer 2

Operating Principle: Bulb

	Degrees of Freedom	Sum of Squares	Mean Square	F Value	Pr(>F)
Test Date	1	0.059	0.059	0.048	0.832
Treatment	1	5.080	5.080	4.134	0.073
Residuals	9	11.060	1.229		

Table E.2: Analysis of variance table for bulb-type sprinklers made by Manufacturer 2 (Sets 2 and 3) used to test the significance of potential test date and treatment effects on sprinkler activation times. Details of the model fit to the data are given in Appendix A. Because the p-values, shown in the column labeled "**Pr(>F)**" are both greater than the chosen significance level of $\alpha = 0.05$, the conclusion is that neither test date nor treatment have any statistically significant effect on the sprinkler activation times. Note, the p-values reported here apply only to the results for Manufacturer 2 for bulb-type sprinklers and have not been adjusted for the additional tests carried out to assess the performance of sprinklers based on other operating principles or made by other manufacturers.

ANOVA Output for Manufacturer 3

Operating Principle: Bulb

	Degrees of Freedom	Sum of Squares	Mean Square	F Value	Pr(>F)
Test Date	1	4.333	4.333	15.561	0.001
Treatment	2	3.136	1.568	5.631	0.013
Residuals	17	4.734	0.279		

Table E.3: Analysis of variance table for bulb-type sprinklers made by Manufacturer 3 (Sets 1, 2, and 3) used to test the significance of potential test date and treatment effects on sprinkler activation times. Details of the model fit to the data are given in Appendix A. Because the p-values, shown in the column labeled "**Pr(>F)**" are both less than the chosen significance level of $\alpha = 0.05$, the conclusion is that both test date and treatment have statistically significant effects on the sprinkler activation times. Note, the p-values reported here apply only to the results for Manufacturer 3 for bulb-type sprinklers and have not been adjusted for the additional tests carried out to assess the performance of sprinklers based on other operating principles or made by other manufacturers. Follow-up multiple comparison analyses were performed to determine exactly how these effects impact the sprinkler activation times, with the results shown on the next page in Table E.4.

Multiple Comparisons Output for Manufacturer 3

Operating Principle: Bulb

Tukey Multiple Comparisons of Means with 95 % Family-Wise Confidence Level

	Difference in Treatment Means	Expanded Uncertainty	Lower Confidence Bound	Upper Confidence Bound
1/31 - 1/28	0.91	0.49	0.42	1.40

	Difference in Treatment Means	Expanded Uncertainty	Lower Confidence Bound	Upper Confidence Bound
Field - Aged	0.36	0.71	-0.36	1.07
New - Aged	1.00	0.78	0.22	1.79
New - Field	0.65	0.71	-0.07	1.36

Table E.4: Multiple comparisons analysis for bulb-type sprinklers made by Manufacturer 3 (Sets 1, 2, and 3) used to compare different factor effects on sprinkler activation times. Further details on the multiple comparisons procedure used are given in Appendix A. Each row in these tables gives a confidence interval for the pairwise difference in treatment levels indicated. The interval for test date is at the 95 % confidence level. The fact that the confidence interval for the difference in test dates does not contain the value 0 confirms the statistical significance of this factor. The confidence interval indicates that the tests made on 1/31/2011 were approximately 0.91 s longer, on average, than the tests made on 1/28/2011 and that the true difference is likely to lie somewhere between 0.42 s and 1.40 s. The three intervals for different treatment comparisons together will simultaneously capture the three true differences in treatment effects with 95 % confidence. Based on these results, the only significant difference between treatments is between new and aged sprinklers, since that is the only interval that does not contain the value 0. This analysis shows that new sprinklers activate approximately 1 s slower than aged sprinklers, on average, and that the true difference most likely lies in the range from 0.22 s to 1.79 s. Differences between the other treatment pairs are smaller and are not significant at this confidence level. The expanded uncertainties for the other two treatment pairs are smaller than for the difference in new and aged sprinkler activation times because more sprinklers taken from the field were measured than new or aged sprinklers.

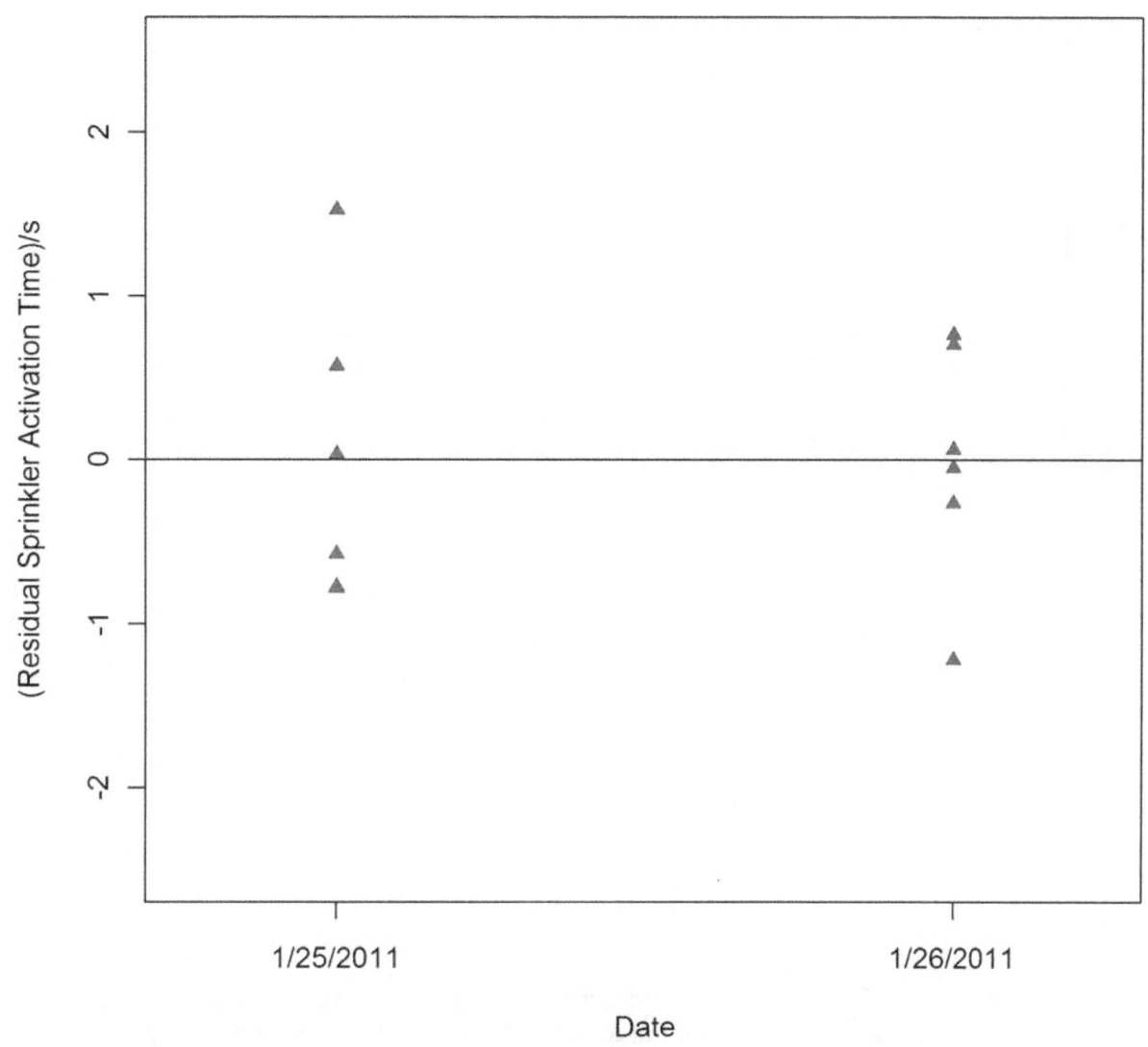

Figure F.1: Residuals from the fit of the model described in Appendix A to sprinkler activation times for fusible-type sprinklers made by Manufacturer 1 (Sets 2 and 3) plotted versus sprinkler test date. This plot indicates that any significant test date effects have been successfully accounted for by the model since the residuals for each test date are centered on a mean value of 0 and display essentially the same level of random variation.

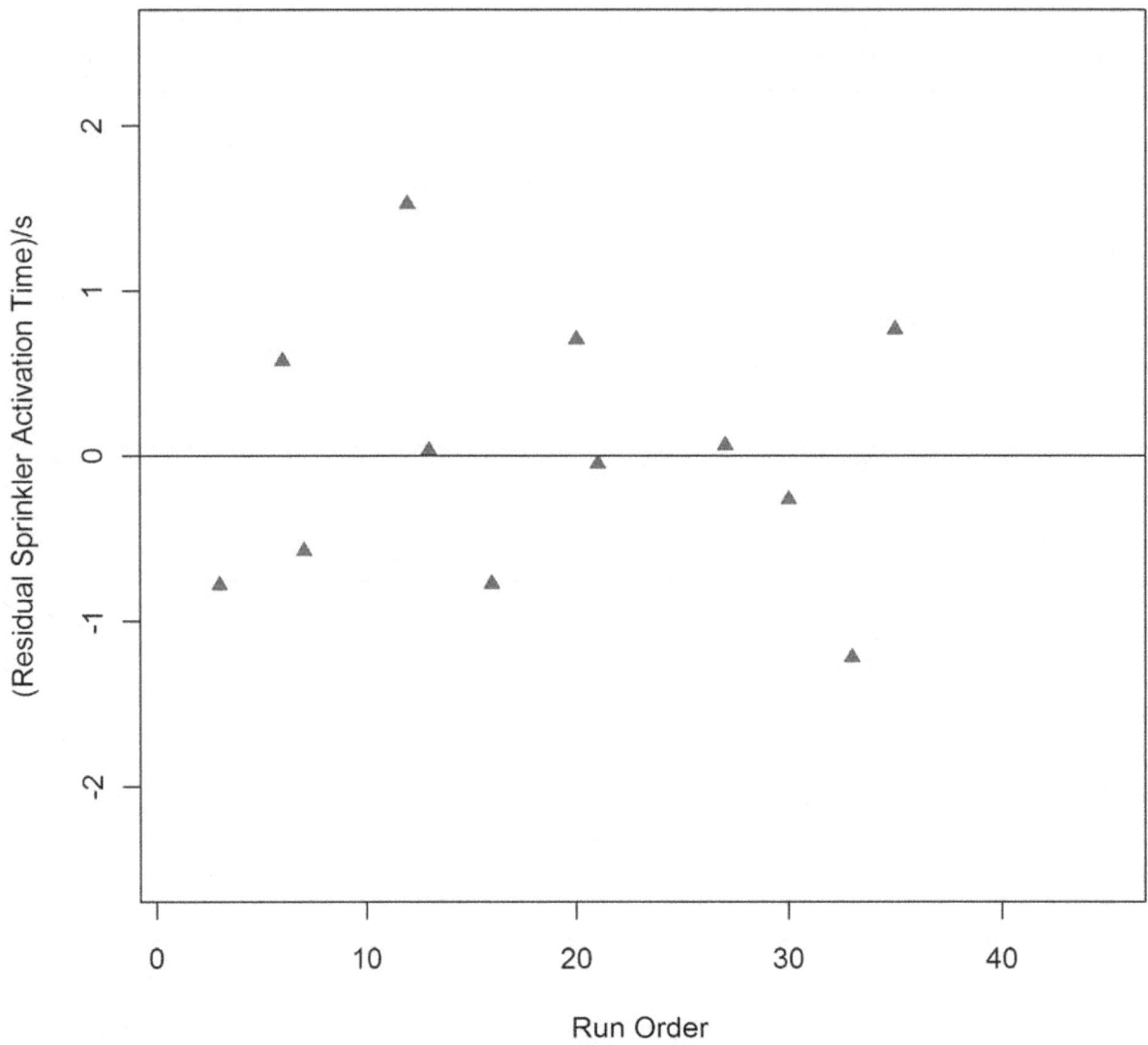

Figure F.2: Residuals from the fit of the model described in Appendix A to sprinkler activation times for fusible-type sprinklers made by Manufacturer 1 (Sets 2 and 3) plotted versus run order. This plot does not identify any drift in the measurements over time since the residuals appear to be randomly scattered around a mean value of 0 and display essentially the same level of random variation across all runs over time.

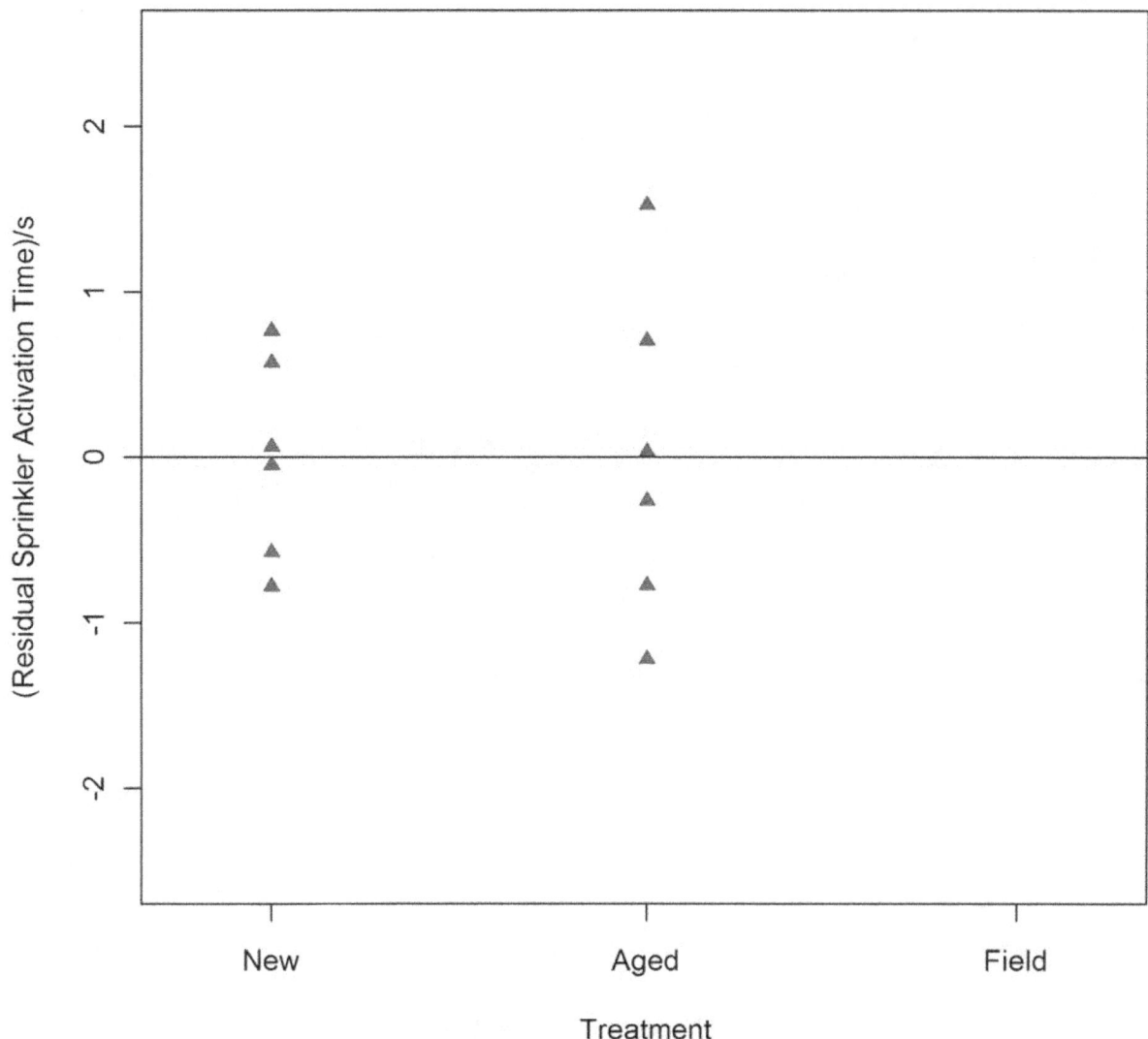

Figure F.3: Residuals from the fit of the model described in Appendix A to sprinkler activation times for fusible-type sprinklers made by Manufacturer 1 (Sets 2 and 3) plotted versus treatment. This plot indicates that any significant treatment effects have been successfully accounted for by the model since the residuals for each test date are centered on a mean value of 0 and display essentially the same level of random variation. No sprinklers of this type and manufacturer were sampled from the field.

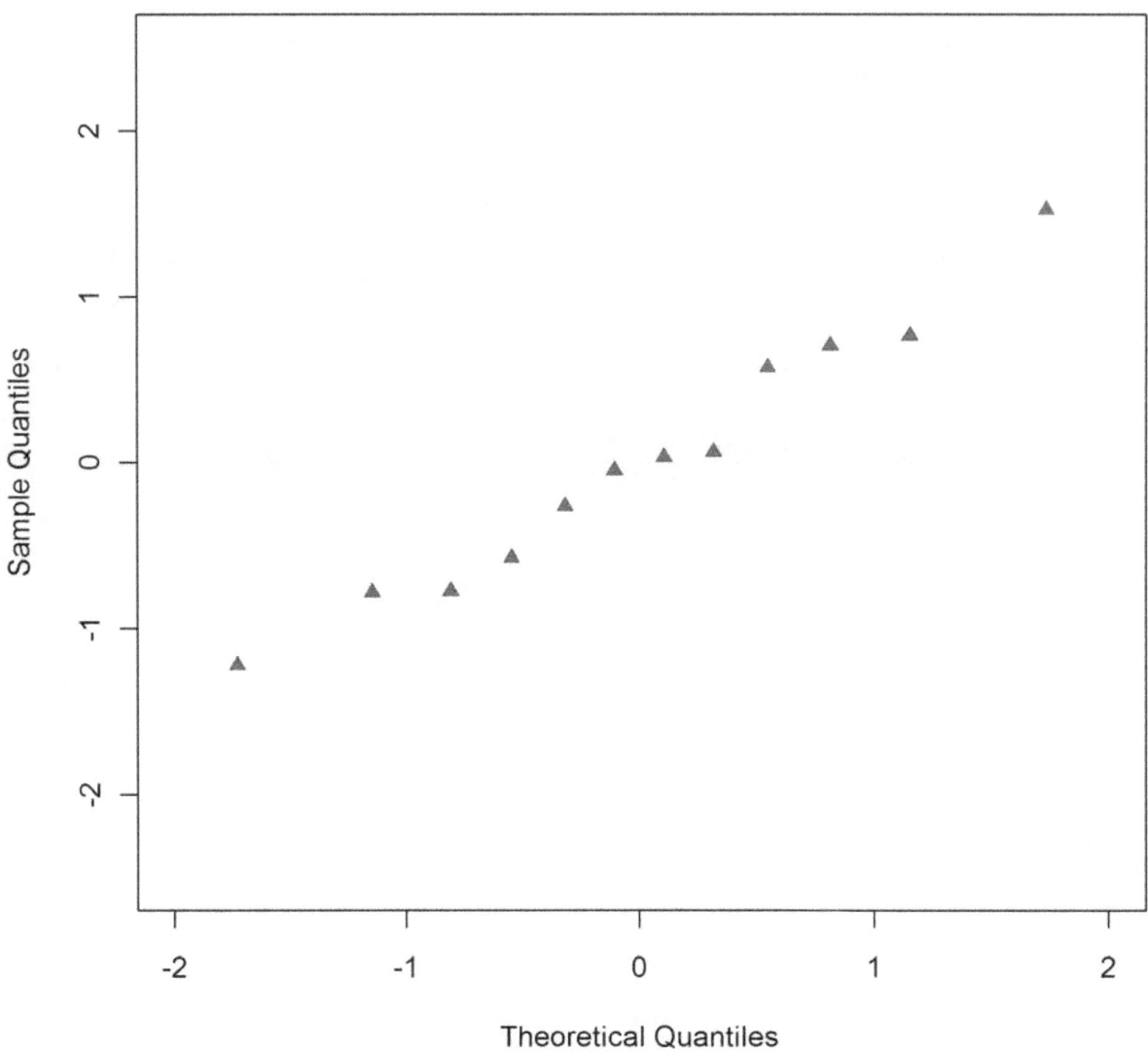

Figure F.4: Normal probability plot of the residuals from the fit of the model described in Appendix A to sprinkler activation times for fusible-type sprinklers made by Manufacturer 1 (Sets 2 and 3). The fact that the residuals essentially fall along a straight line when plotted versus theoretical quantiles from the standard normal distribution suggests that the random errors follow a distribution that reasonably approximates the assumed normal distribution.

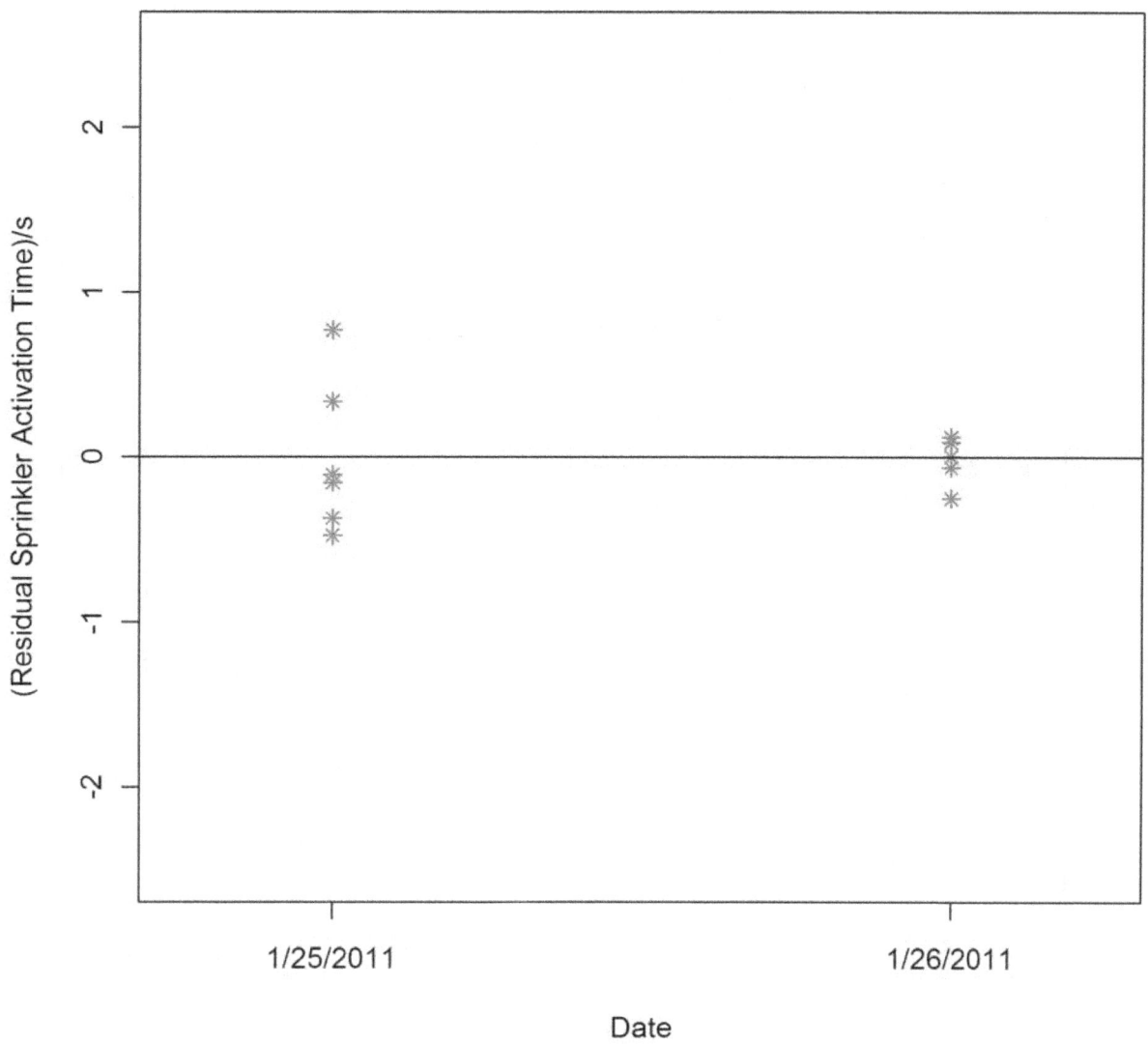

Figure F.5: Residuals from the fit of the model described in Appendix A to sprinkler activation times for fusible-type sprinklers made by Manufacturer 2 (Sets 2 and 3) plotted versus sprinkler test date. This plot indicates that any significant test date effects have been successfully accounted for by the model since the residuals for each test date are centered on a mean value of 0 and display essentially the same level of random variation.

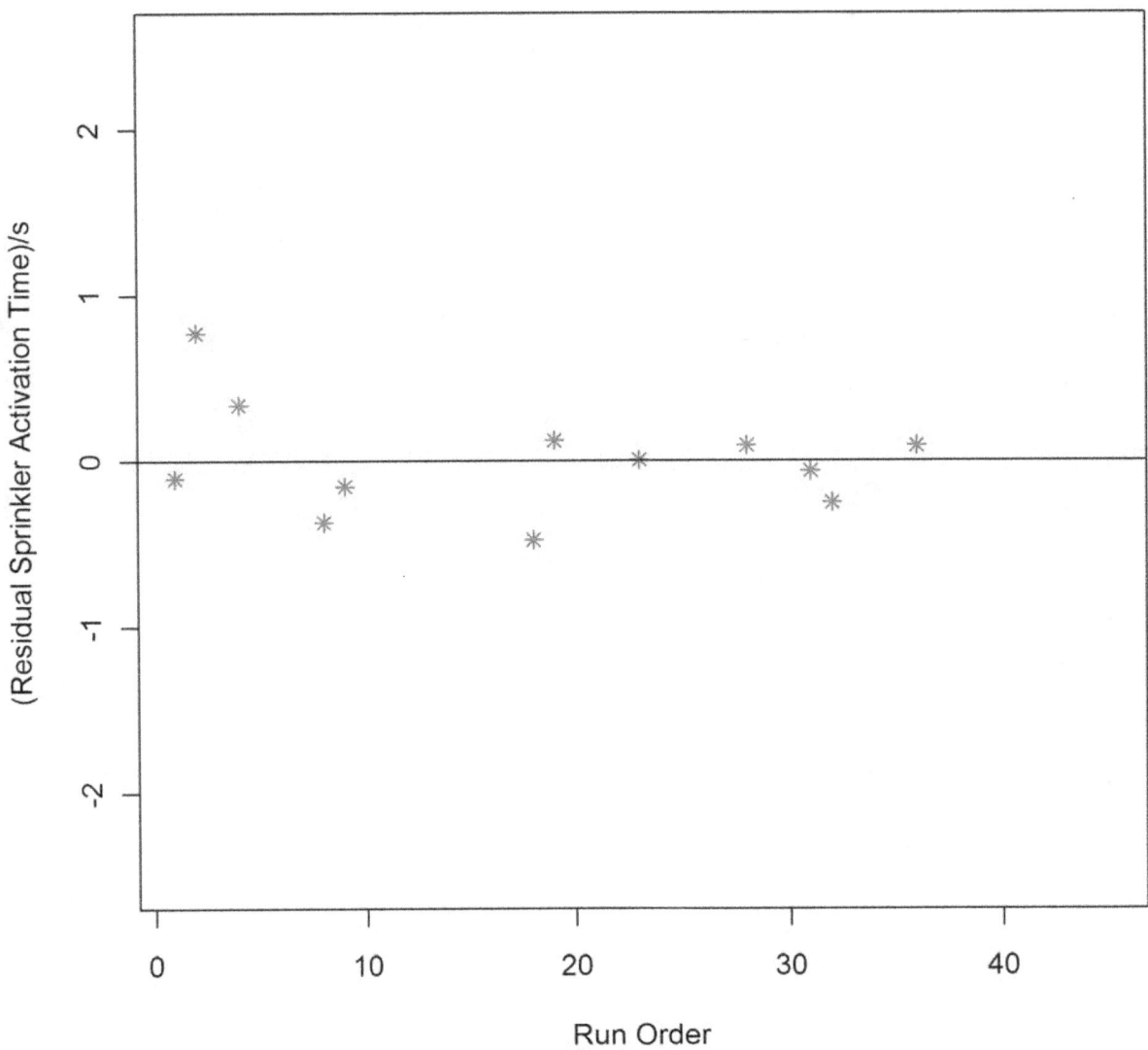

Figure F.6: Residuals from the fit of the model described in Appendix A to sprinkler activation times for fusible-type sprinklers made by Manufacturer 2 (Sets 2 and 3) plotted versus run order. This plot does not identify any drift in the measurements over time since the residuals appear to be randomly scattered around a mean value of 0 and display essentially the same level of random variation across all runs over time.

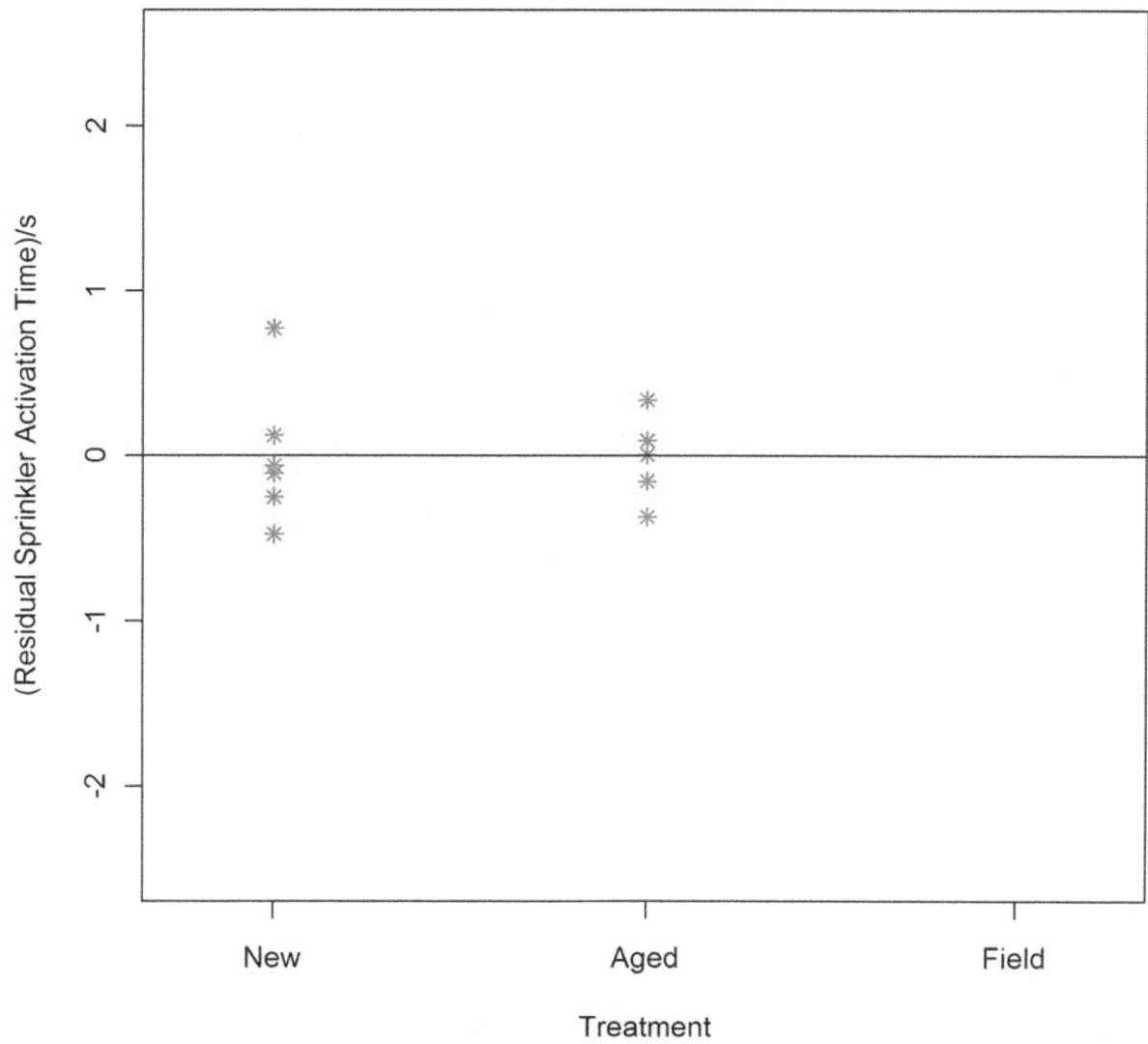

Figure F.7: Residuals from the fit of the model described in Appendix A to sprinkler activation times for fusible-type sprinklers made by Manufacturer 2 (Sets 2 and 3) plotted versus treatment. This plot indicates that any significant treatment effects have been successfully accounted for by the model since the residuals for each test date are centered on a mean value of 0 and display essentially the same level of random variation. No sprinklers of this type and manufacturer were sampled from the field.

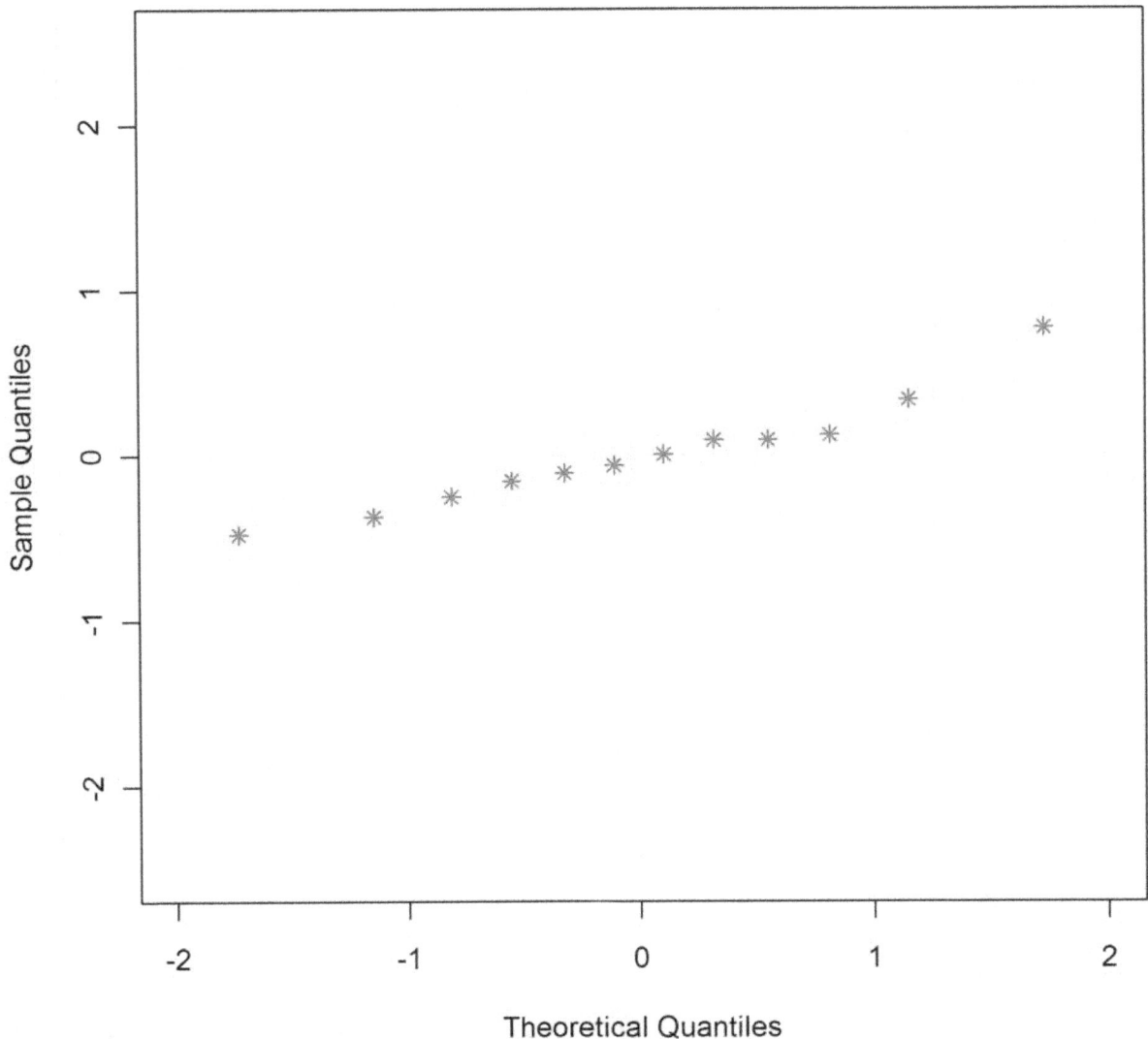

Figure F.8: Normal probability plot of the residuals from the fit of the model described in Appendix A to sprinkler activation times for fusible-type sprinklers made by Manufacturer 2 (Sets 2 and 3). The fact that the residuals essentially fall along a straight line when plotted versus theoretical quantiles from the standard normal distribution suggests that the random errors follow a distribution that reasonably approximates the assumed normal distribution.

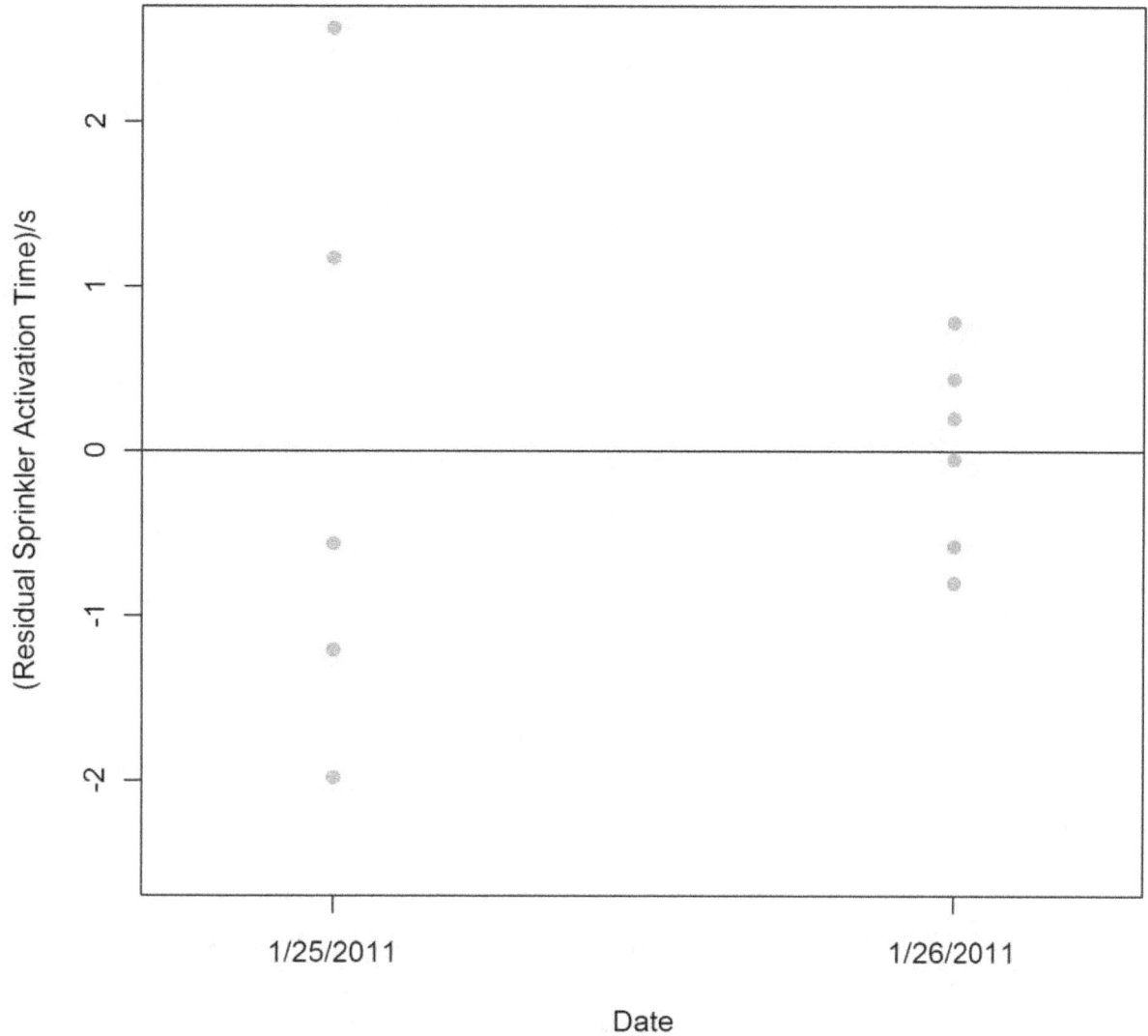

Figure F.9: Residuals from the fit of the model described in Appendix A to sprinkler activation times for fusible-type sprinklers made by Manufacturer 3 (Sets 2 and 3) plotted versus sprinkler test date. This plot indicates that any significant test date effects have been successfully accounted for by the model since the residuals for each test date are centered on a mean value of 0 and display essentially the same level of random variation.

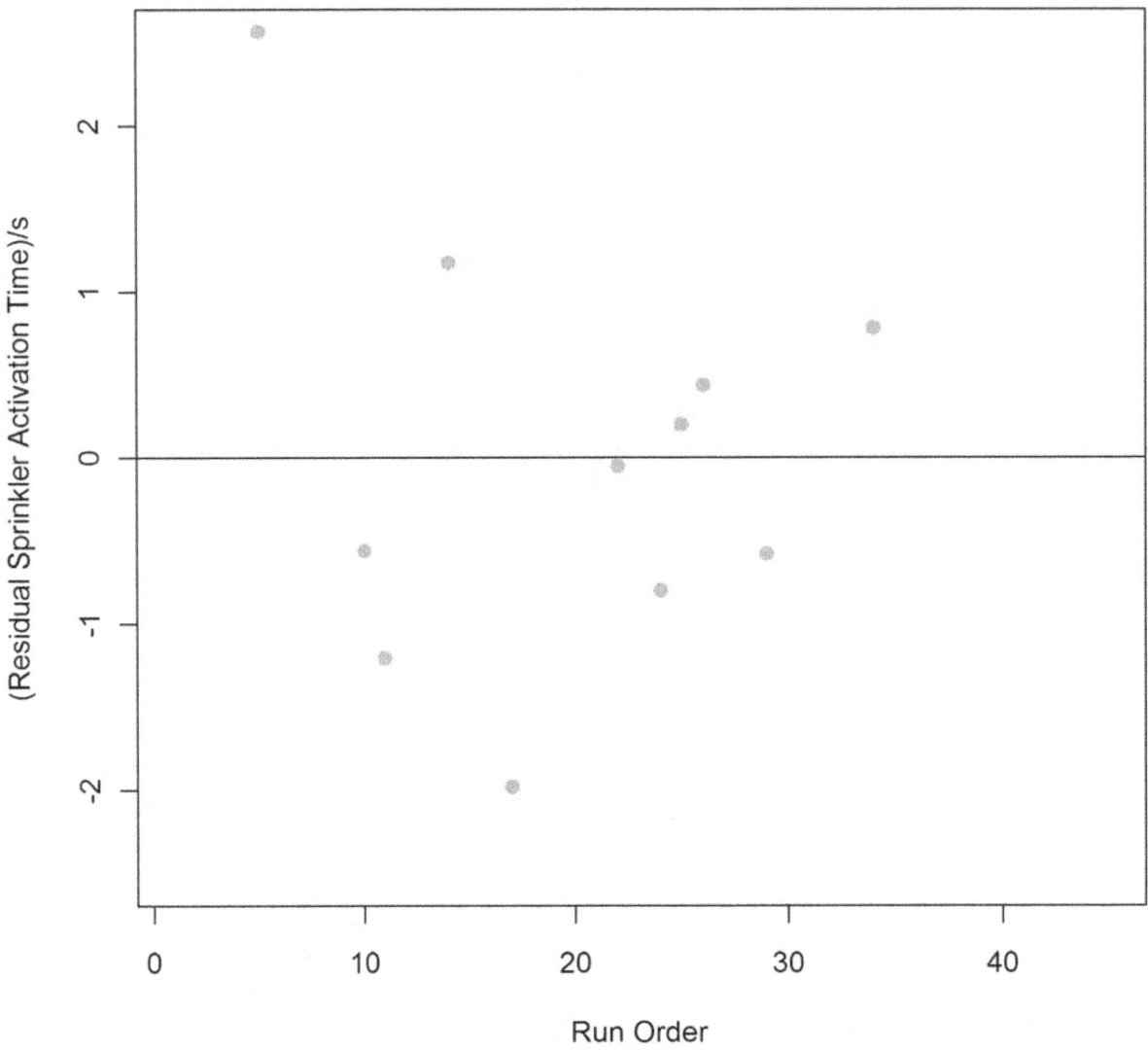

Figure F.10: Residuals from the fit of the model described in Appendix A to sprinkler activation times for fusible-type sprinklers made by Manufacturer 3 (Sets 2 and 3) plotted versus run order. This plot does not identify any drift in the measurements over time since the residuals appear to be randomly scattered around a mean value of 0 and display essentially the same level of random variation across all runs over time.

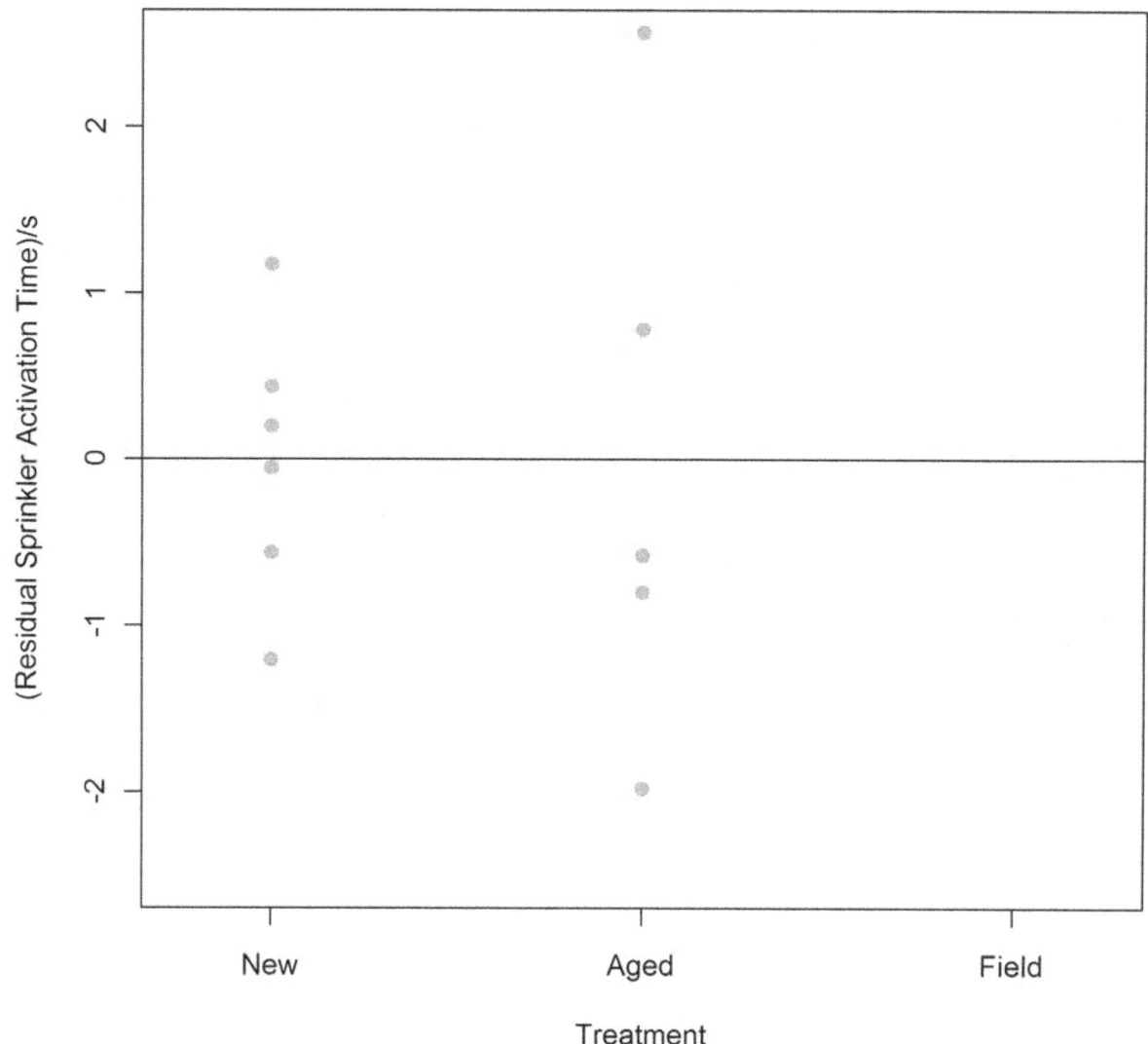

Figure F.11: Residuals from the fit of the model described in Appendix A to sprinkler activation times for fusible-type sprinklers made by Manufacturer 3 (Sets 2 and 3) plotted versus treatment. This plot indicates that any significant treatment effects have been successfully accounted for by the model since the residuals for each test date are centered on a mean value of 0 and display essentially the same level of random variation.

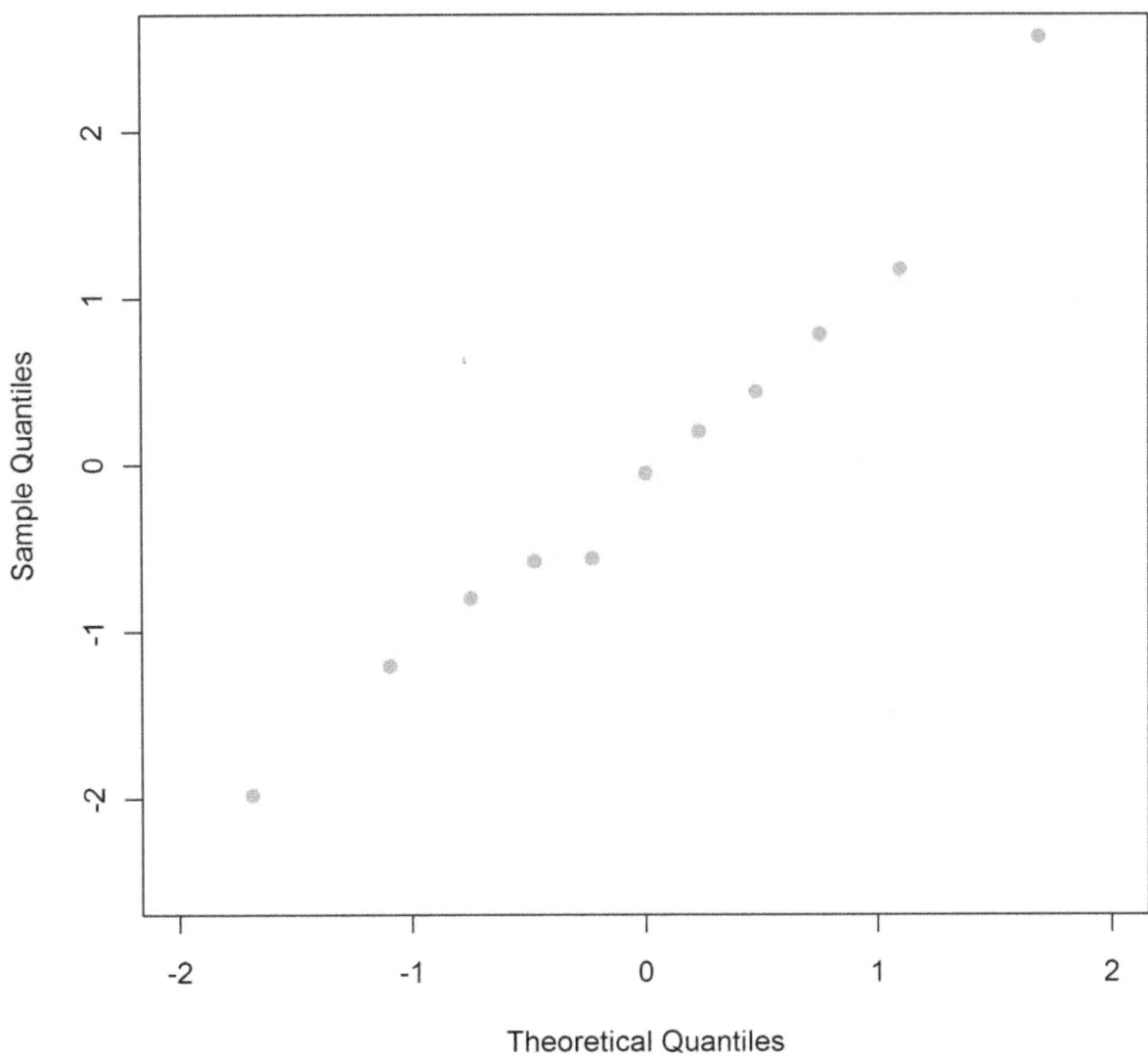

Figure F.12: Normal probability plot of the residuals from the fit of the model described in Appendix A to sprinkler activation times for fusible-type sprinklers made by Manufacturer 3 (Sets 2 and 3). The fact that the residuals essentially fall along a straight line when plotted versus theoretical quantiles from the standard normal distribution suggests that the random errors follow a distribution that reasonably approximates the assumed normal distribution.

ANOVA Output for Manufacturer 1

Operating Principle: Fusible

	Degrees of Freedom	Sum of Squares	Mean Square	F Value	Pr(>F)
Test Date	1	0.526	0.526	0.692	0.427
Treatment	1	2.815	2.815	3.705	0.086
Residuals	9	6.838	0.760		

Table F.1: Analysis of variance table for fusible-type sprinklers made by Manufacturer 1 (Sets 2 and 3) used to test the significance of potential test date and treatment effects on sprinkler activation times. Details of the model fit to the data are given in Appendix A. Because the p-values, shown in the column labeled "**Pr(>F)**" are both greater than the chosen significance level of $\alpha = 0.05$, the conclusion is that neither test date nor treatment have any statistically significant effect on the sprinkler activation times. Note, the p-values reported here apply only to the results for Manufacturer 1 for fusible-type sprinklers and have not been adjusted for the additional tests carried out to assess the performance of sprinklers based on other operating principles or made by other manufacturers.

ANOVA Output for Manufacturer 2

Operating Principle: Fusible

	Degrees of Freedom	Sum of Squares	Mean Square	F Value	Pr(>F)
Test Date	1	0.139	0.139	1.034	0.336
Treatment	1	0.431	0.431	3.214	0.107
Residuals	9	1.207	0.134		

Table F.2: Analysis of variance table for fusible-type sprinklers made by Manufacturer 2 (Sets 2 and 3) used to test the significance of potential test date and treatment effects on sprinkler activation times. Details of the model fit to the data are given in Appendix A. Because the p-values, shown in the column labeled "**Pr(>F)**" are both greater than the chosen significance level of $\alpha = 0.05$, the conclusion is that neither test date nor treatment have any statistically significant effect on the sprinkler activation times. Note, the p-values reported here apply only to the results for Manufacturer 2 for fusible-type sprinklers and have not been adjusted for the additional tests carried out to assess the performance of sprinklers based on other operating principles or made by other manufacturers.

ANOVA Output for Manufacturer 3

Operating Principle: Fusible

	Degrees of Freedom	Sum of Squares	Mean Square	F Value	Pr(>F)
Test Date	1	0.161	0.161	0.083	0.780
Treatment	1	5.442	5.442	2.815	0.132
Residuals	8	15.468	1.934		

Table F.3: Analysis of variance table for fusible-type sprinklers made by Manufacturer 3 (Sets 2 and 3) used to test the significance of potential test date and treatment effects on sprinkler activation times. Details of the model fit to the data are given in Appendix A. Because the p-values, shown in the column labeled "**Pr(>F)**" are both greater than the chosen significance level of $\alpha = 0.05$, the conclusion is that neither test date nor treatment have any statistically significant effect on the sprinkler activation times. Note, the p-values reported here apply only to the results for Manufacturer 3 for fusible-type sprinklers and have not been adjusted for the additional tests carried out to assess the performance of sprinklers based on other operating principles or made by other manufacturers..

This page intentionally left blank.

REFERENCES

1. Averill, J.D., Gann, R.G., Murphy, D.C., and Guthrie, W.F., "Performance of New and Aged Residential Fire Smoke Alarms." NIST Technical Note 1691. National Institute of Standards and Technology, Gaithersburg, MD, April 2011.

2. http://www.cpsc.gov/info/drywall.

3. Environmental Health & Engineering, Inc. "Final Report on an Indoor Environmental Quality Assessment of Residences Containing Chinese Drywall." EH&E Report 16512, submitted to the Consumer Product Safety Commission, Bethesda, MD, January 28, 2010.

4. Maddalena, R., Marion, R., Moya M., and Apte, M.G., "Small-Chamber Measurements of Chemical Specific Emission Factors for Drywall." Report Number: LBNL-3986E, Ernest Orlando Lawrence Berkeley National Laboratory, Berkeley, CA, October 2010.

5. Abbott, W., "The Development and Performance Characteristics of Mixed Flowing Gas Test Environment," *IEEE Transactions on Components, Hybrids, and Manufacturing Technology* **11**, 22-35, (1988).

6. Glass, S. J., Mowry, C. D., and Sorensen, N. R. "Report on Accelerated Corrosion Studies of Electrical Components." Sandia Report SAND2011-1539. Sandia National Laboratories, Albuquerque, NM, March 2011.

7. *2009 International Residential Code for One- and Two-Family Dwellings.* International Code Council, Inc., Falls Church, VA.

8. *2009 International Building Code.* International Code Council, Inc., Falls Church, VA.

9. *UL 1626: Standard for Residential Sprinklers for Fire-Protection Service.* Underwriters Laboratories. Northbrook, IL, (2008).

10. *NFPA 13: Standard for the Installation of Sprinkler Systems.* National Fire Protection Association. Quincy, MA, (2007).

11. *NFPA 13D: Standard for the Installation of Sprinkler Systems in One- and Two-Family Dwellings and Manufactured Homes.* National Fire Protection Association. Quincy, MA, (2007).

12. *NFPA 13R: Standard for the Installation of Sprinkler Systems in Residential Occupancies up to and Including Four Stories in Height.* National Fire Protection Association, Quincy, MA, (2007).

13. *UL 199: Standard for Automatic Sprinklers for Fire-Protection Service.* Underwriters Laboratories. Northbrook, IL, (2005).

14. Heskestad, G., and Bill Jr., R. G., "Quantification of Thermal Responsiveness of Automatic Sprinklers Including Conduction Effects." *Fire Safety Journal*, 14 (1988) 113 – 125.

15. Madrzykowski, D., "The Effect of Recessed Sprinkler Installation on Sprinkler Activation Time and Prediction." Master's Thesis, University of Maryland, College Park, MD, 1993.

16. *Guide to the Expression of Uncertainty in Measurement.* International Standards Organization, ISO/TAG 4/WG 1, 2008, Geneva, http://www.bipm.org/utils/common/documents/jcgm/JCGM_100_2008_E.pdf.

17. *NFPA 25: Standard for the Inspection, Testing, and Maintenance of Water-Based Fire Protection Systems.* National Fire Protection Association, Quincy, MA, (2008).

18. Mee, R.W., "Simultaneous Tolerance Intervals for Normal Populations with Common Variance." *Technometrics*, Vol. 32, No. 1, 1990.

19. Montgomery, D.C., *Design and Analysis of Experiments.* 2nd ed., John Wiley and Sons, New York, 1984.

20. Miller, R.G., *Simultaneous Statistical Inference.* Springer, Berlin, 1981.

21. NIST/SEMATECH e-Handbook of Statistical Methods. http://www.itl.nist.gov/div898/handbook/prc/section4/prc471.htm, April 13, 2011.

22. R Development Core Team, *R: A Language and Environment for Statistical Computing.* R Foundation for Statistical Computing, Vienna, Austria, 2011, http://www.R-project.org.